SANDSTONE CENTER OF THE WORLD
Images & Stories of Quarry Life in Amherst, South Amherst, & Lorain County, Ohio

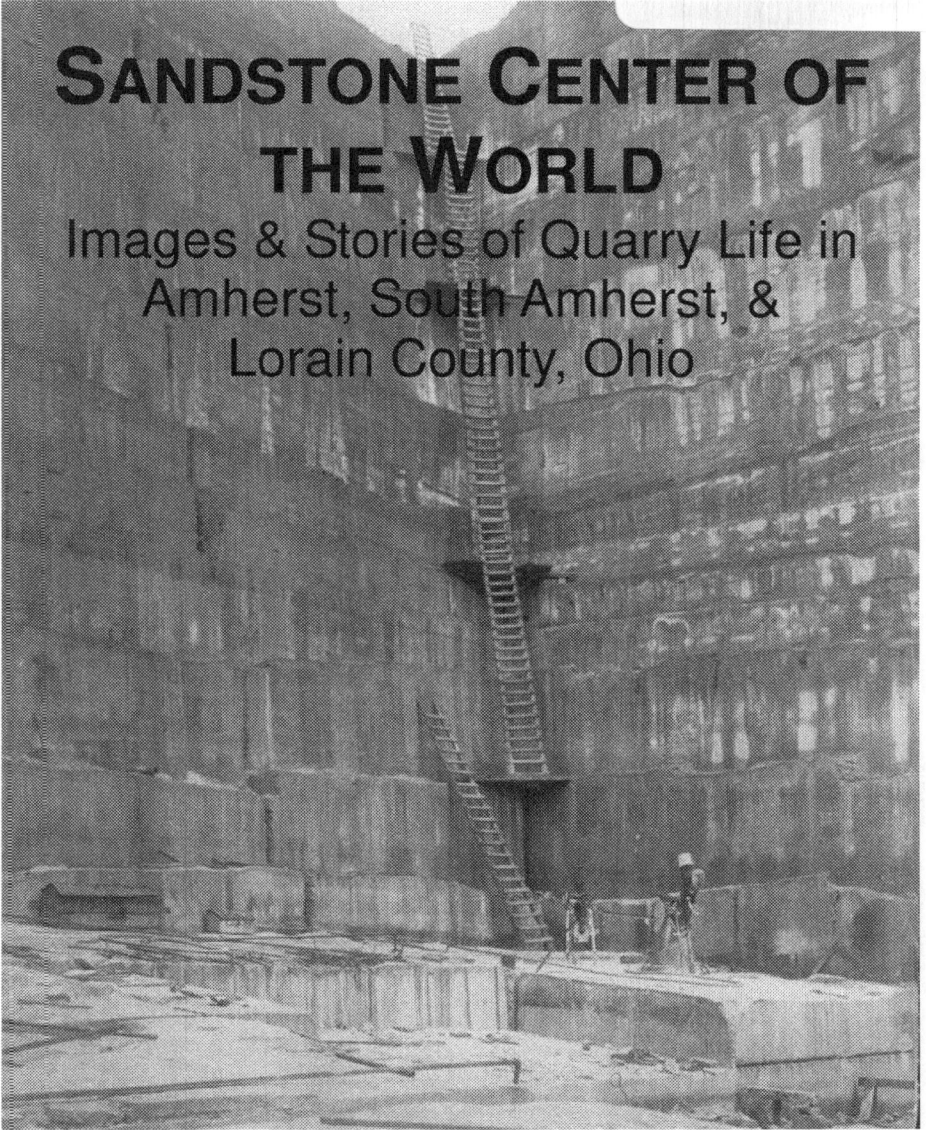

James A. Hieb
Foreword by Sally Cornwell

www.quarrytown.net

Sandstone Center of the World

Images & Stories of Quarry Life in
Amherst, South Amherst, &
Lorain County, Ohio

Quarrytown.net Publishing
109 Squires Court
South Amherst, Ohio 44001
www.quarrytown.net
tripp42@aol.com
ISBN 978-0-6151-4653-9

A section of the #7 Quarry with a small crew of men working in July 1959 (Bill & Mary Louise Back photograph.)

DEDICATION...

I would like to dedicate this book to my wife Christa, our two rapidly growing boys, Kyle and Chad, and Christa's grandfather Foyster Matlock whose story about working at the Cleveland Quarries inspired this book.

AMAZING QUARRIES...

During its heyday, the quarries in Lorain County were a sight to behold:

I was hardly prepared to realize the vast magnitude of the work going on here. The stone annually handled is simply enormous. In nearly all these quarries work was being vigorously pushed although it was late in the closing season. It was cheery and pleasant to hear the click, click of the pick, chisel and drill, as I went from quarry to quarry.

> - W.W. Williams, History of Lorain County, Ohio—1879.

It is, however, in Cuyahoga, Erie and Lorain counties that the Berea grit is found in its most admirable condition, and that quarried at Amherst, in Lorain county, is taken to represent it at its best....The production and sale of this valuable material employ a small army of men, and the industry has grown to be one of the heavy business interests in Northern Ohio.

> - Manufacturer and Builder—1887.

The product of the "Buckeye" and "Amherst" quarries is grey and buff sandstone of the highest quality.

> - Mine and Quarry—1906.

No place more dramatically portrays man's victory over Nature in more definite terms than this deep, canyon-like, man-made quarry reaching into the earth.

> - The Cleveland Quarries—circa 1940.

Today, one cannot look into the depths of these quarries without a feeling of awe and wonder. Mother Nature has been slowly reclaiming them, but look closely at the grooves from the augers the quarry workers drilled to cut loose huge blocks of stone. If the wind is just right and you listen intently, you may hear the sounds of past generations extracting something beautiful from the earth and the master carvers transforming sandstone into works of art.

> - Sally Cornwell—2006.

Cleveland Quarries Company—A unit of Amherst Stone Company (AMST), primary business is in the quarrying and fabricating of sandstone.

Lorain County—A county located in the northeast corner of the state of Ohio, United States. Named by early settler Herman Ely. Home to over 50 quarries—most now abandoned.

Quarry—The location of an operation where a deposit of stone is extracted from the earth through an open pit or underground mine.

Sandstone—A sedimentary rock consisting of usually quartz sand united by some cement (as silica or calcium carbonate).

Sandstone Center of the World—Official designation received by the City of **Amherst** in 2003 by the Ohio Bicentennial Commission.

South Amherst—Village in Lorain County that, along with Amherst, was a prominent home to the largest sandstone quarries in the region. Also home of the author.

THE BUCKEYE FIRESTONE COCKTAIL

1/3 Grand Marnier
1/6 Orange Juice
1/3 Gordon's Gin
1/6 Lemon Juice

Shake well in cracked ice and serve in cocktail glass

In 2006 the Cleveland Quarries Company donated a collection of documents to the Amherst Historical Society that included original dated blueprints, pictures, maps, and countless other records. One of the many items included amongst the hundreds of items donated included this recipe card (printed on business card stock) which was likely used by the company management when entertaining potential clients.

SANDSTONE

FROM THE

Amherst, Ohio, Sandstone Quarries——

THE SANDSTONE CENTER OF THE WORLD

"BUCKEYE GRAY"		"BIRMINGHAM GRAY"
"AMHERST BUFF"	SANDSTONE	"BIRMINGHAM WARMTONE BUFF"
"AMHERST MULTI-COLORED MAHOGANY"		"BIRMINGHAM BUFF"

THE OHIO QUARRIES COMPANY
UNION TRUST BUILDING   CLEVELAND, OHIO.

Before the Cleveland Quarries Company emerged on April 17, 1929, there were dozens of stone companies that played a role in making the Amherst region the "Sandstone Center of the World." Worthington & Sons and The Ohio Quarries Company are just two quarry companies featured in this book.

CONTENTS

> Amherst Historical Society
> Lorain County Map
> Cleveland Quarries Company
> No. 7 Quarry Winter Scene
> Sandstone Center of the World Stone Marker

Quarry workers stop to pose for this picture circa 1920. A closer look discloses rugged faces and worn work boots, yet the picture also captures the tools of time. Picks, shovels, chains, and hammers showcase the backbreaking nature of the work (photograph courtesy of Fay Ott).

THE CLEVELAND QUARRIES COMPANY
NO. 7 QUARRY, SOUTH AMHERST, OHIO

Penny postcard of the No. 7 quarry (Amherst Historical Society collection).

FOREWORD

BY SALLY CORNWELL

Amherst, Ohio has been known as "The Sandstone Center of the World" since 1910 and this title became official in 2003 through the Ohio Historical Society. The Amherst, Amherst Township and South Amherst area is located on top of a huge deposit of billion-year-old sandstone and whose history surrounds the quarrying of this stone. A fortune has been made out of these deposits of the earth. This sandstone has proved to be an important economic blessing to our early settlers and is the foundation of the county's rich heritage.

Amherst sandstone was found to be ideal for making grindstones. When the region's early settlers saw the outcroppings of stone, the poor timber and pasturage, they felt this was a serious blemish to this land. By 1847, sandstone from this region was known to have excellent sharpening qualities. This was the beginning of the ideal Amherst grindstone. Sandstone was then used as a prized building material. Some of the most admired architecture in the world contains sandstone from the Amherst area: The John Hancock Mutual Life Insurance building in Boston, Buffalo City Hall, The Hockey Hall of Fame in Toronto, the Auditorium at Long Beach, California, and the Cornell University in New York, just to name a few.

An awe-inspiring example of Amherst sandstone and craftsmanship may be seen on the Hope Memorial Bridge in Cleveland, Ohio. This bridge was originally called the Lorain-Carnegie Bridge. It was renamed in 1983 in honor of Bob Hope's father, William Henry Hope. William Hope was a stonemason and a master carver. The men who carved this bridge were artists in the real sense of the word.

During the early 1900's, the Amherst sandstone quarry industry was the largest of its kind in the world. At that time, the Ohio Geological Survey said, "Amherst stone is commended by the following qualities which it possesses in an unusual degree: durability, strength, color and texture."

Today, one cannot look into the depths of these quarries without a feeling of awe and wonder. Mother Nature has been slowly reclaiming them, but look closely at the grooves from the augers the quarry workers drilled to cut loose huge blocks of stone. If the wind is just right and you listen intently, you may hear the sounds of past generations extracting something beautiful from the earth, and the master carvers transforming sandstone into works of art.

Over the years, the Amherst Historical Society (AHS) had amassed a small collection of articles and pictures about the quarries in the Amherst area. During the research and documentation needed for the City of Amherst to officially receive the designation of the "Sandstone Center of the World," I had the pleasure of working with Russ Ciphers, Sr. and the Cleveland Quarries Company, who provided me access to many documents in their archives. Without their cooperation and support, the 2004 designation might not have been possible. At that time it was my hope that these records might one day receive professional preservation so that they could be enjoyed by generations to come.

Then in 2006, a small miracle occurred. The Cleveland Quarries Company donated this collection of documents that included original dated blueprints, U.S. and Canadian patents, labor union records, deeds, pictures, maps, and countless other records—many of which date from the early 1880s. Youngstown State University's Department of History professors, proficient in industrial archival collections, called the collection "a historical treasure." AHS is now working on preservation of these documents, as well as the creation of a traveling exhibit and collection of oral histories.

The impact of this collection will no doubt spurn additional research and countless new perspectives of quarry life in the Amherst region. I was pleased to meet Jim Hieb, who independently had been working on a book about quarries in Lorain County, especially South Amherst, as a tribute to his wife's grandfather, Foyster Matlock, who retired from the Cleveland Quarries. The timing could not have been more perfect. Jim has blended his own version of the quarry story, aided by many of the new AHS resources, with an attention to detail and a fresh perspective.

We are fortunate that this book was created and I'm proud to add it to my reference library. I hope you enjoy it too!

-**Sally Cornwell**

Sally is employed at the Lorain County Clerk of Courts. She served as the assistant to the Amherst Mayor from 1996 to 2003 and was instrumental in assisting the city to receive the "Sandstone Center of the World" designation during the State of Ohio's Bicentennial. Sally is active in the American Legion Auxiliary, Amherst Democrat Women's Club, and a number of other Amherst civic organization.

Penny postcards from the early 1910s. Workers pose as the sandstone blocks are prepared to be cut into slabs by giant gang saws.

The Ohio Quarries Company's sign promotes the famed "Buckeye" gray sandstone and tours of the Buckeye Quarry and Mills. A dominant player in the local market from 1903 to 1919, consolidated with other firms in 1919 and finally merged to form the Cleveland Quarries in 1929 (Amherst Historical Society collection).

THE QUARRIES, AMHERST, OHIO.

Penny postcards from the 1910s. Several quarry photos were used for postcards during this time period (courtesy of Fay Ott).

INTRODUCTION

Since the early 1800s the quarrying of sandstone in and around Amherst (previously North Amherst) and South Amherst, Ohio, has led to the region to become known as the "**Sandstone Center of the World**." At the height of its hey-day, several dozen quarries were active and over 2,000 individuals were employed locally. While sandstone operations are now only a fraction of what they once were, its legacy and tradition can be found rather easily. Grindstones, the original sandstone product envisioned by John Baldwin, dot the local countryside. Many prominent structures in the region, built using Amherst or Berea Sandstone, still stand. Several businesses, such as the Quarry Café in downtown Amherst, use the term "quarry" in their firm's name. Quarry Road, a widely-used county road, was moved several times to make room for expansions in the size of the adjacent quarries.

For the past few years, county, village, and township officials have been discussing a proposal to turn much of the land around the Cleveland Quarries Company into a high-profile development that would include a golf course, hotels, a gated-residential community, and much more. The company is still looking for investors to purchase the property and if successful, the landscape of the area will change, yet the deep heritage the quarries brought to the region will remain.

The history and continued popularity of sandstone in the region is the primary reason the Cleveland Quarries Company, a unit of American Stone Industries (AMST), announced in 2007 that it will be moving its fabrication, warehouse, and distribution operations of the stone from the current South Amherst location to a facility in Vermilion, Ohio. The total investment will be over three million dollars in an existing building and state-of-the-art machinery. This relocation and expansion positions the company for an extremely bright future.

Sandstone Center of the World

The title "Sandstone Center of the World" first caught public attention when an Amherst businessman, O.H. Baker, used this slogan on his business stationary in 1910. Amherst government officials later erected a sign on the town square with this slogan at the corner of Main Street and Milan Avenue. Some forty years later this sign, and

Current sign at the corners of Main Street and Milan Avenue in downtown Amherst, Ohio. The first sign was erected in 1915 (Jim Hieb collection).

the claim made by Amherst residents, angered residents of South Amherst as to which community was the rightful heir to this title. In 1955, the sign was temporary removed. It was later replaced after officials at the Cleveland Quarries confirmed that sufficient amounts of sandstone were derived from quarries in both communities.

While there are other Ohio communities that argue a significant role in our nations sandstone quarry history (for example, Berea, Ohio), Amherst has capitalized the most on the use of the slogan. The sandstone found in both Amherst and South Amherst are part of a larger deposit that stretches from the southeastern corner of Ashtabula County, westward into Erie County, and south to Adams County along the Ohio River. Many counties exist in this region, yet the best sandstone is from Lorain and Cuyahoga counties. This bed of sandstone is part of the "Berea Geological Formation" yet most of the stone coming out of Lorain County has been referred to as simply "Amherst Sandstone."

Signs designating the "Sandstone Center of the World"
A number of signs in Amherst promote the slogan and designation of the city as the sandstone capital.

The Ohio Historical Marker that resides outside the Amherst City Hall identifies, commemorates, and honors the important people, places and events surrounding the sandstone industry. The marker states:

"AMHERST—Sandstone Center of the World"
The City of Amherst was founded in 1811. Beginning in 1847, Amherst developed and prospered around the sandstone industry and its associated quarries. This sandstone proved to be an important economic blessing to our early settlers and is

Historical Marker outside of the Amherst City Hall (Jim Hieb collection).

the foundation of Amherst's existence. Amherst sandstone shows lines of stratification or bedding when exposed in sections. Its natural beauty is enhanced by a virtual spectrum of rich and unique colors including deep reds, browns, yellows, and shades of gray. Amherst sandstone is well known for its quality, durability, and rich texture and has been utilized across our nation and throughout the world. Amherst is literally and figuratively built upon a rock, which extends deep in the earth.

Erected in 2003, the marker was sponsored by the Ohio Bicentennial Commission, Longaberger Company, Amherst Committee for the Ohio Bicentennial, and The Ohio Historical Society. The marker is designated with the number 14-47, which means it was the 14th Ohio Historical Marker placed in Lorain County.

Awe-Inspiring Quarries

There is something awe-inspiring about quarries. Just imagine how the stone was brought out of the ground to make its way to become the building blocks for ancient and modern buildings. How was marble brought down from the mountains around Carrara, Italy? How did

---

the ancient Egyptians move the vast stone blocks to make the world famous pyramids? What methods were used to extract sandstone in Ohio?

The **South Amherst quarries** grew deeper and deeper over time and employed derricks and pulleys to bring the stone out from the quarry's bottom for production. This method was complex and dangerous. By the early 20th century, the Buckeye Quarry was believed to be the deepest in the world at 240 feet deep—today there are several quarries that are deeper.

Many quarries that opened in the 20th century have employed a much different method to extract the stone than their older Ohio counterparts. A good example is a granite quarry located near Milbank, South Dakota. As the picture depicts, you'll see that the quarry design has taken into account the use of a drive-in quarry using motorized vehicles to remove the stone blocks for processing from almost every layer of the quarry. Compare this method to the one used by the South Amherst quarries and many like it that were opened in the 19th century. One quickly sees how the differences used to extract the stone likely played a role in the safety of these quarrying activities.

Example of a drive-in quarry in South Dakota. Notice each segment of the quarry is accessible by motorized vehicle vs. a derrick system of extracting the stone blocks from the quarry bottom (Dakota Granite Company photograph.)

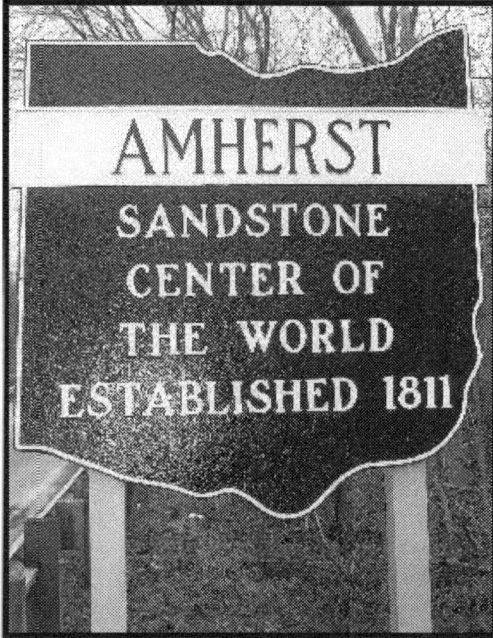

Visitors and residents are reminded that Amherst is the "Sandstone Center of the World." The signs were installed at five of the city entrances through the joint efforts of the Ohio Historical Society, City of Amherst, Ohio Bicentennial Committee, Main Street Amherst, Amherst Beautification Committee, and the Amherst American Legion Elmer Johnson Post 118 in 2005 (Jim Hieb collection).

The Amherst Rotary club used sandstone from the Cleveland Quarries to fabricate the "Rotary Wheel" that resides in Rotary Park in front of Amherst Hospital on Cleveland Street (Jim Hieb collection).

Stories of those Who Toiled in the Quarries

The quarry boom in Lorain County began with the opening of a quarry in Brownhelm Township. Yet the inspiration of this book began with the stories of a different sort; they came from my wife's grandfather, Foyster Matlock. His story is like many told by those who toiled to make a living in the quarries. It is one of hard work, supporting a family, and building friendships to last a lifetime. I close the introduction of this book with his story. You will find additional reflections on quarry life throughout this book.

Foyster Matlock spent nearly 30 years with the Cleveland Quarries working in a variety of laborer capacities, primarily in the quarry's fabrication department. Previously he had served during World War II and was a coal miner from Trosper, Kentucky and when work had become scarce in the Trosper area he looked for other opportunities. In 1950 at the age of 36, he temporarily left his wife and eight children to come to South Amherst looking for that opportunity. He was not the only one with a Kentucky background to look to the Amherst area, as his father-in-law Robert Frederick (retired in 1953 as a fireman at the Cleveland Quarries) and others had preceded him. The automobile, steel, shipbuilding, and quarry industries in Lorain County were all benefactors of hard-working laborers moving north.

His initial employment with the quarries did not last long. While he admits that he initially quit because "he could," his first stint with the quarries only lasted 2-3 months due largely be-

<<< **Foyster Matlock pay stub** from the Cleveland Quarries Company: 45 Regular Hours, 5 Overtime Hours, $81.23 Gross Pay, $1.62 FICA & Withholding Tax, $4.85 deducted for Insurance, $3.00 deducted for union dues, $71.76 Net Pay issued on check #32161 (Jim Hieb collection).

cause he missed his family back in Kentucky. He returned to the quarries soon thereafter, and in 1951 brought his entire family to join him in Ohio. Like others who moved into houses owned by Cleveland Quarries, the cost of rent and electricity was taken directly from his weekly payroll check. The quarry-owned house lacked running water which meant it had to be obtained from a nearby well or from water jugs.

Matlock described the work at the quarries as "down-right labor....all work...no foolishness." In the fabrication building, he and other laborers would load by hand 80-plus pound sandstone pieces that had come down a conveyor belt from the breaking machine onto railroad boxcars all day long.

Admitting that he never had any trouble with his bosses, he did, however, participate in a couple of union strikes during his tenure with the company. He was never an organizer, yet did support the cause of the union by participating on the picket line. The short-lived strikes were organized for the typical labor concerns of wages, safety, and general treatment of the employees. When asked if he held up picket signs and shouted chants about labor concerns, he said, "we just mulled around near the gate (entrance) on Quarry Road." With little fanfare, he spent his time on the picket line with Hurstle Dalton, a long-time friend. By Dalton's family's admission, he was someone who could embellish a story and keep your attention. "He could lie with the best of them, and tell really good ones (stories)" stated Matlock. While on the picket line, his son J. Russell Matlock would drive by and drop off soda pop for his father and Mr. Dalton.

Occasionally business would slow down and the fabrication shop employees would be asked to work a 4-day workweek. To pick up an additional 8 hours to complete his 40 hours, Matlock would volunteer to work in another department. He did this four times during his quarry days working 1 day in the quarry hole, 1 day for the quarry railroad, and 2 days as a stone crusher. While he preferred working in his own department, it was a way to maintain uniform wages for the pay period.

Several of Matlock's sons and sons-in-law worked at the Cleveland Quarries for brief periods of time including time before and after serving our nation in the armed forces during the Vietnam War. Several of Matlock's children, grandchildren, and great-grandchildren still live in the South Amherst area.

Note: Mr. Matlock was 3 weeks shy of his 91st birthday at the time of this interview (2/3/05). He turned 93 shortly before the completion of this book.

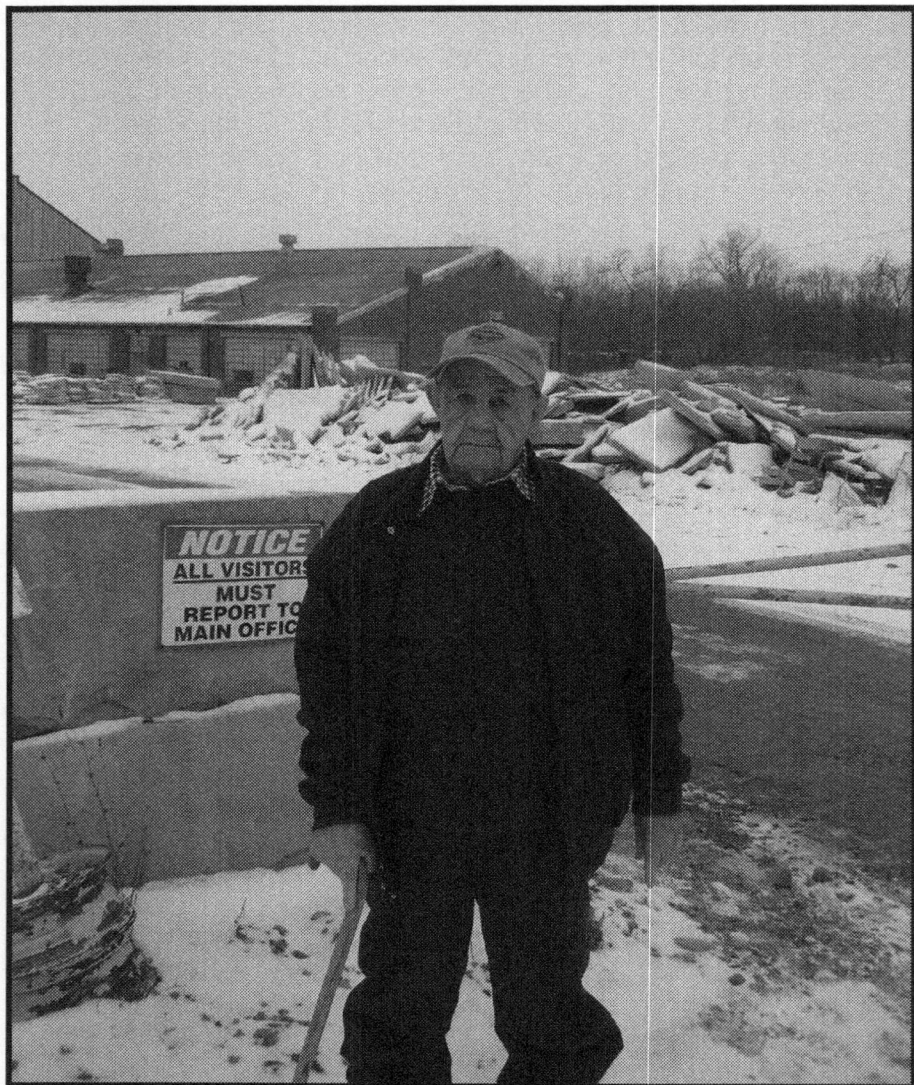

Foyster Matlock stands outside the No. 8 Mill on Quarry Road in January 2007. The building behind him was used as the cut stone department. During Matlock's tenure at the quarries, he lived in two different quarry-owned houses (both a 'stone's throw' from this location). Most of the quarry houses have long been torn down and the streets have been overgrown with trees and brush.

The No. 7 Quarry is directly west; the Kamm Quarry directly north; and the Buckeye Quarry southwest of this site (Jim Hieb collection).

Derrick used in the 1910s (Amherst Historical Society collection).

The famous 1916 Amherst train wreck has been documented in many Amherst historical documents. The overpass, like many others in Amherst, was constructed of Amherst sandstone (photo courtesy of Fay Ott).

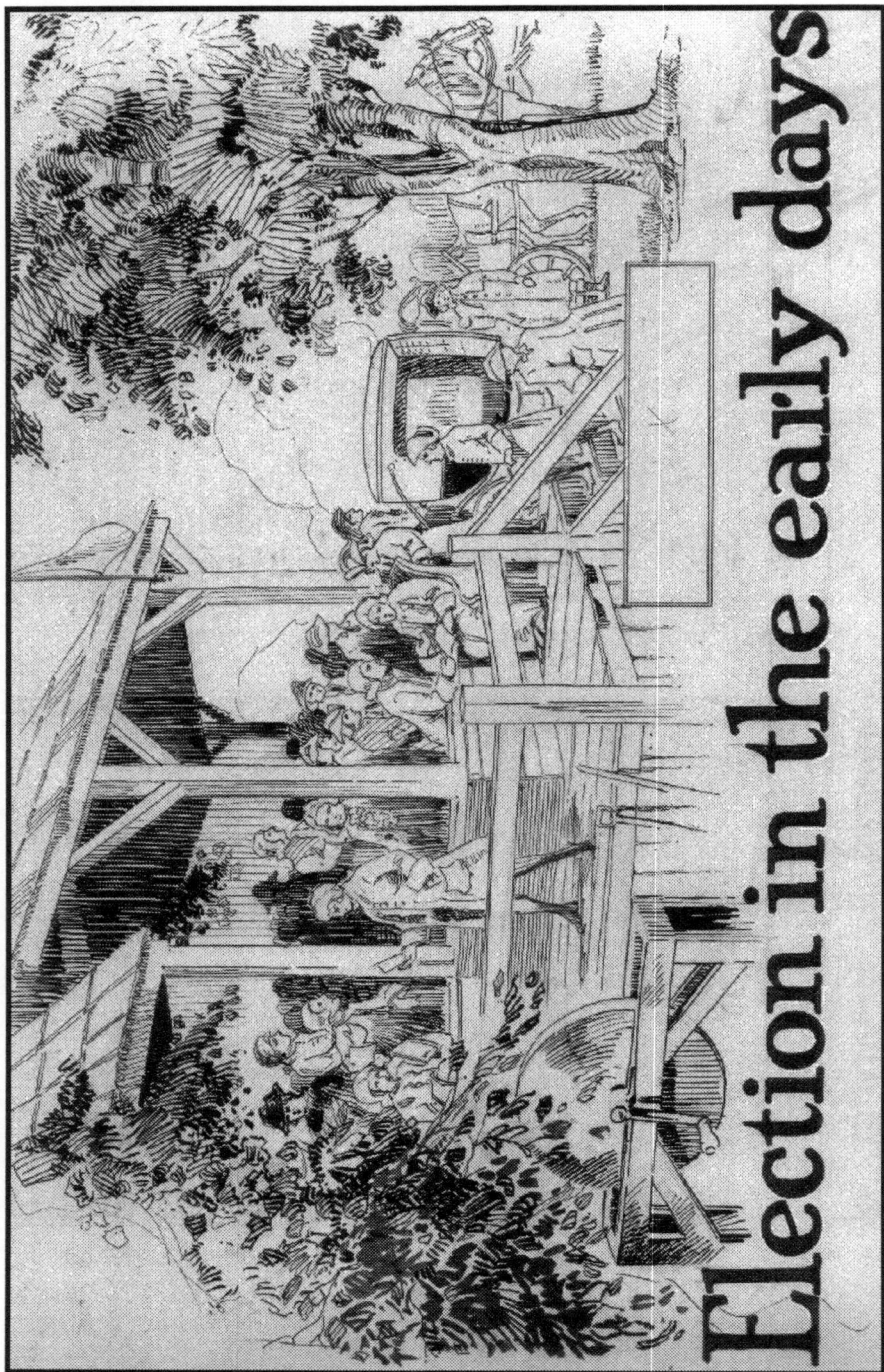

Election in the early days

Sketch used by the Cleveland Quarries highlighting the importance grindstones played in our nation's history—voting wasn't the only reason to gather at the town center (AHS collection).

1 QUARRY HISTORY

Quarry Boom Extends to Lorain County

It is widely accepted that the quarrying boom in the region actually began in the city of Berea in neighboring Cuyahoga County. As this boom extended into Lorain County, dozens of quarries would be in operation over the years. Many of the extensive quarry operations are referenced, yet there were also many smaller quarries located on farms using the stone for local buildings or a neighbor's use.

John Baldwin, Sr.

The name "Berea" in Ohio has a connection to a city, a sandstone formation, and a man named John Baldwin, Sr. Baldwin (1799-1884) and his wife Mary moved from Connecticut to Middleburg Township in Cuyahoga County in 1828. He set out to establish an educational-religious community called Lyceum Village. This effort would fail causing financial hardship.

John Baldwin, Sr.

Baldwin's luck would change when he noticed how easily he could sharpen a knife on the sandstone along the riverbeds of his Middleburg Township property. He soon invented a lathe, powered by water power, to cut slabs of the sandstone into grindstones. Over time, this would become an important industry that would catapult him into a successful business. A quarry boom began as others opened quarries in and around Baldwin's property, as well as the adjacent counties.

Baldwin would use his financial earnings and deeding some of his property, to found Baldwin University (now Baldwin-Wallace College) in 1845, which was one of the first colleges in Ohio to admit students without regard to race or gender.

Baldwin also played a role in the naming of a village near his home as "Berea" in memory of the biblical reference of Paul and Silas as depicted in Acts:

*"And the brethern immediately sent away Paul and Silas by night into **Berea** who coming thither went into the synagogue of the Jews. These were more noble than those in Thessalonica, in that they received the word with all readiness of mind, and searched the scriptures daily, whether those things were so."*

-Acts 17:10-11

The "Berea" reference has also been applied to the sandstone formation that stretches through much of Ohio. Berea sandstone would quickly be recognized as the premier building material used throughout the country. The Berea community would gain the designation of "Grindstone City," as well as the "Grindstone Capital of the World." While Baldwin did not reside in Lorain County, he made a significant impact on the region's rich sandstone history.

Chronological History of Quarry Activity

Like any industry, there was an evolution of events and companies that played a major role in the development of the region's sandstone industry. This chronicle of the many full-scale operations will reveal several names that were prominent including Worthington, Nichols (sometimes spelled with two "l"s in referenced documents), and Clough, amongst others. Again, as the City of Amherst has capitalized on the designation of the "Sandstone Center of the World," the entire region played a role. Residents of South Amherst, Brownhelm and Amherst townships, Kipton, Grafton, Elyria, Black River (Lorain), and others area communities played a role in the rich heritage of the region's sandstone industry.

1828

John Baldwin, Sr., Founder of Baldwin University (now Baldwin-Wallace College), Berea, Ohio, moved from Connecticut to Middleburg Township, Cuyahoga County, Ohio.

1830

Amherst population was 552, Brownhelm Township 388, and South Amherst is unincorporated.

1842

Mr. Baldwin credited with the idea of utilizing local stone to **create the first lathe-made grindstone** using water power from a local mill. It would not be until 1873 that Baldwin would form a company to quarry sandstone—the Baldwin Quarry Company.

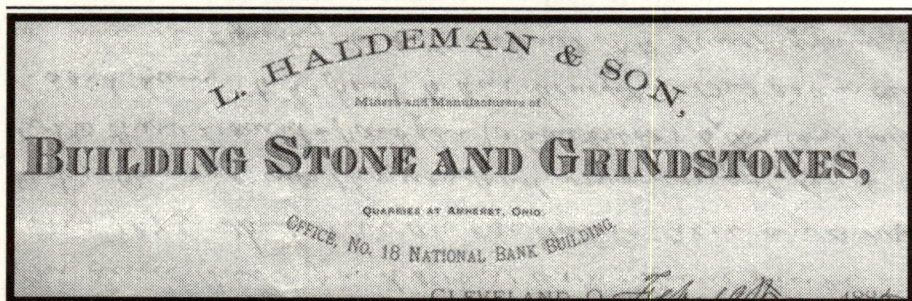

L. HALDEMAN & SON,
Miners and Manufacturers of
BUILDING STONE AND GRINDSTONES,
Quarries at Amherst, Ohio
OFFICE, No. 18 NATIONAL BANK BUILDING
CLEVELAND, O.

1847
Henry Warner opened the Brownhelm Quarry (the first known quarry in Lorain County) on a tract of land he purchased for $600. He is believed to have shipped the first stone out from this area.

1848
Wm. Wilson & Company opened a quarry about one mile east of the Village of North Amherst.

Sylvester Silsby manufactured grindstones from the John Elliott Quarry.

1852
Baxter Clough opened a quarry about two miles west of the Village of North Amherst near the L.S. & M.S. Railroad. This quarry later became known as Clough's Middle Quarry.

1853
John Worthington, a contractor from Toronto Canada, purchased the Brownhelm Quarry for $6000. The quarry land contained "Buff" Sandstone denoting the color of the stone (learn more about John Worthington below).

Black River Stone Company is believed to have quarried sandstone in Carlisle Township (just south of Elyria) for the First Presbyterian Church (Old Stone Church) on Public Square in downtown Cleveland.

1863
William James purchased a quarry about one mile northeast of North Amherst and produced building stone for local trade. This quarry is likely the John Elliott Quarry which he worked for fifteen years.

The Wilson & Hughes Stone Company opened quarries on two parcels of land in North Amherst. Initially the stone was hauled to Black River (now Lorain) or North Amherst. In 1868 the Lake Shore

---

Grindstone City, Michigan
John Worthington would later own quarries in Grindstone City, Michigan. At the corner of Copeland and Rouse Roads, in Grindstone City, is a huge grindstone donated to the town by the Cleveland Quarries Co., in honor of the pioneers who spent their lives working at the grindstone industry there. Several sale ads for grindstones in the Amherst Historical Society archives reference stone from Grindstone City being sold by Amherst-based stone companies.

& Michigan Southern Company extended a railroad spur into the quarries.

1865
Foster Brothers took over the William James Quarry and operated it for approximately three years. It was then sold to George E. Hall & Company.

1867
James Nichols purchased a quarry adjacent to his farm that was owned by Joseph Barber. This quarry would later be known as the Nichols Quarry (or the Old East Quarry). Later, this quarry was simply known as the #9 quarry of the Cleveland Stone Company.

1868
George E. Hall and Company built and operated the first stone saw mill in the area. The plant was called The Amherst Stone Works.

The **Grafton Stone Company** opened a quarry in Grafton, Ohio, called the Black River Quarry.

1869
R. P. Wilson Company opened the No. 6 Gray Canyon Quarry. This quarry is reputed to be the largest single sandstone quarry in the world. The land, totaling approximately 40 acres, was purchased from John Siep.

The **Clough Stone Company** opened two additional quarries: one north of the Clough Middle Quarry, known as the Clough North Quarry, and another south of the Clough Middle Quarry became known as the Clough South Quarry. A railroad siding of small gauge track was constructed connecting these quarries to Lake Erie, thus enabling shipments via boat, as well as rail. This track ran downhill to the lake and the flatbed cars would be pulled back to the quarry by horses, which caused the track to become known as the "Pony Trail." All three quarries would later be sold to the Cleveland Stone Company and become known as the No. 8 Quarry.

Once part of Lake Erie
The Amherst quarries in Lorain County are located in a series of ledges, which were once the shore cliffs of Lake Erie. The elevated position produces a free drainage, and the stones have been traversed by atmospheric water to such a degree that all processes of oxidation which are possible have been nearly completed. The elevation also facilitates extraction.

-Manufacturer and Builder, May 1887

1871

John Worthington & Sons purchased the quarry and mill from George E. Hall and Company. They also constructed a railroad track from Brownhelm Station to Lake Erie in Vermilion.

H.E. Mussey & Company of Elyria opened a quarry in Elyria and became a large producer of grindstones.

1870s

A number of people purchased small tracts of land and opened quarries. The frequency of new quarries being opened has been compared to the Gold Rush of 1847.

John Hancock Mutual Life Insurance Company, Boston, Massachusetts

As many as 50 or more now-abandoned quarries are thought to be hidden by dense brush and trees. Some of the companies operating during this time where: The Ohio Stone Company, Grove Quarry, Butler Quarry, Nichol and Miller Company, Wilson Hughes Stone Company, Turkey Rock Quarry, Haldeman & Son, Clough Stone Company, and the Amherst Stone Company.

Amherst population was 2,482, Brownhelm Township 1,461, and South Amherst was unincorporated.

Following the great Chicago fire in 1871, extensive amounts of Amherst sandstone were used in the rebuilding efforts. The beauty and durability of sandstone from the Amherst region, coupled with easy access transporting stone on the Great Lakes, was a natural for Chicago.

Stonington Lodge Free & Accepted Masons Lodge #503

On October 21, 1875, the charter for the Masonic lodge in Amherst was obtained. Is it coincidence that the name "Stonington" was applied this new lodge given the scope of the sandstone industry in the Amherst area? Take note that the list of lodge officers included none other than J.M. Worthington and James Nichols.

1872
Amherst Stone Company opened its quarry in the fall of 1872 on property that was known as the old Quigley property behind the present-day IGA. Like other stone companies, they also operated grindstone lathes and a saw mill. The Quigley House, part of the Amherst Historical Society properties, is built from stone from this quarry.

1874
James Nichol and the Clough Stone Company acquired twenty-four acres of land adjacent to the north side of the L. Haldeman & Son quarry. Their quarry would later be merged into the No. 6 (Gray Canyon) quarry.

W.H. Bryant opened the Turkey Rock Quarry—so called because tracks of turkeys were found at the site. The grindstones made from this site were said to rival the famous Wickersley Stones of England and the stone would be known as American Wickersley.

1875
James Nichol and Dr. Cross purchased the James Wood farm in Florence Township in Erie County and opened a quarry. This was the first Wakeman Buff Stone Quarry. One year later, James Nichol

formed a partnership with Dudley Miller, known as Nicholl & Miller, and assumed operations of the Wakeman Quarry.

1876
Atlantic Stone Company opened the Nickel Plate Quarry at Nickel Plate, Ohio (approximately four miles east of Oberlin).

1877
John Worthington died. His business endeavors continued by his two sons (James M. and George H.) using the name Worthington & Sons.

1879
Jos. Barber opened a quarry on land adjoining the L. Haldeman & Son quarry to the east.

1880
James Nichol opened and operated a quarry on his own property located about one mile east of North Amherst. Portions of this quarry are visible from the current State Route 2.

The **Ohio Building Stone Company** of Cleveland incorporated and operated as a selling agent for the building stone produced by various Amherst stone operators.

The **Ohio Grindstone Company** of Cleveland incorporated and operated as a selling agent for grindstones produced by various Amherst stone operators.

1883
Clough Stone Company purchased a tract of land and began quarry operations adjacent to the Atlantic Stone Company in Nickel Plate. They also organized the Oberlin and LaGrange Railroad Company. In 1884 they constructed approximately four miles of railroad to the L.S. & M.S. Railroad siding in Oberlin for the purpose of transporting stone from the Nickel Plate Quarries to customers.

1884
Worthington & Son purchased the Jos. Barber Quarry, which was

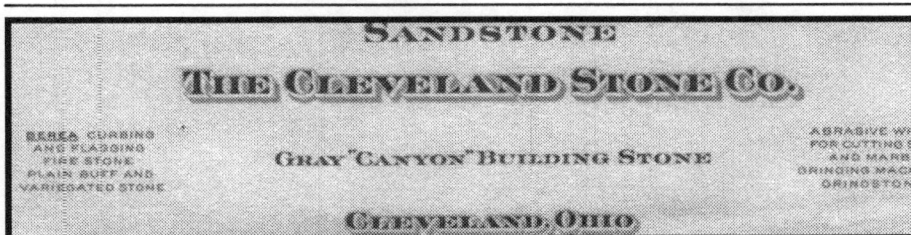

SANDSTONE

THE CLEVELAND STONE CO.

BEREA CURBING
AND FLAGGING
FIRE STONE
PLAIN BUFF AND
VARIEGATED STONE

GRAY "CANYON" BUILDING STONE

ABRASIVE WH
FOR CUTTING S
AND MARB
GRINDING MAC
GRINDSTON

CLEVELAND, OHIO

later absorbed by the Cleveland Stone Company and merged in the No. 6 quarry.

John Baldwin Sr. died.

Lorain Times Newspaper reports that **Wilson Quarries** will "lie idle for at least the next two seasons. This will throw out of employment about one hundred and fifty employees—though one-half of this number can find employment at other quarries."

1886
Clough, Haldeman & Atlantic Stone Company was incorporated. The new company served as a selling agent for products produced by the Clough Stone Company, L. Haldeman & Sons, and the Atlantic Stone Company.

July 26, 1886
The **Cleveland Stone Company** was incorporated. This company included several existing quarries and businesses including: Berea Stone Company, J. McDermott & Company, Clough Stone Company, Worthington & Sons, James Nichol, Nichol & Miller, Columbia Stone Company, the Ohio Building Stone Company, Ohio Grindstone Company, and the Berea & Huron Stone Company. The Cleveland Stone Company would later purchase the interests of L. Haldeman & Son, the Atlantic Stone Company, the Lake Huron Stone Company, as well as interest in other smaller companies. These consolidations were largely through the efforts of George H. Worthington.

1887
The **Elyria Stone Company** was formed by Charles T. Ely and his brother. They opened and operated a quarry near Grafton.

1888
Malone Stone Company opened a quarry near the No. 6 Quarry.

1890
Several companies consolidate operations under the name **Ohio**

Stone Company. These companies include: The Ohio Stone Company, Haldeman Stone Company, and Wilson & Hughes Stone Company.

1892
Wm. H. Bryant opened a quarry south of the No. 6 Quarry. Bryant also invented a machine known as the "Bryant Channeler" for cutting lime and sandstone.

1893
Mussey Stone Company of Elyria became incorporated and opened a quarry south of the No. 6 Quarry. This property was later acquired by the Cleveland Stone Company and became known as the No. 7 Quarry.

West of Lorain County in Berlin Heights (Erie County), the **Baillie Stone Company** was incorporated by G. A. Baillie and others.

1894
Kipton Stone Company in Kipton, Ohio, was incorporated.

1895
Forest City Stone Company incorporated and began quarry operations in Euclid (Cuyahoga County) and Columbia Station (Lorain County).

1897
Cleveland Stone Company absorbed the operations of The Mussey Stone Company, The Malone Stone Company, and the Kipton Stone Company.

1903
Ohio Quarries Company was organized by John R. Walsh, a large Bedford, Indiana quarry operator. It promoted its product as "Buckeye" Gray Sandstone. Note: the Bedford, Indiana area is still famous for the quarrying of Indiana Limestone.

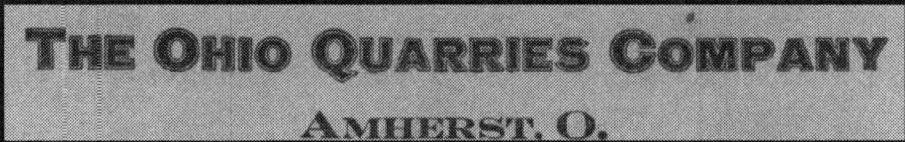

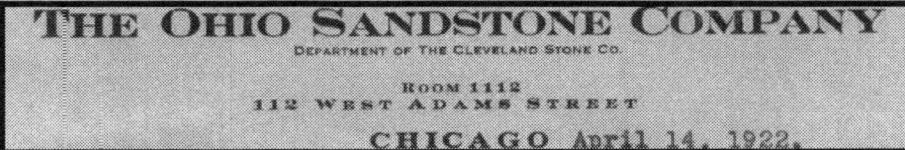

THE OHIO QUARRIES COMPANY

AMHERST, O.

THE OHIO SANDSTONE COMPANY

DEPARTMENT OF THE CLEVELAND STONE CO.

ROOM 1112
112 WEST ADAMS STREET

CHICAGO April 14, 1922.

1904
The Independent Stone Company was incorporated and operated a quarry about two miles west of North Amherst.

1908
The Breakwater Company of Cleveland assumed operations of the Independent Stone Company. After operating the quarry for one year, it was abandoned.

1909
Quarry No. 9, located on the south side of North Ridge Road, was closed. This quarry was also known as the Old East Quarry (formerly the Nichols Quarry).

1911
Blum and Debridge Cut Stone Company moved from Cleveland to the Amherst area and began operations. This company specialized in cut stone with the use of highly trained stonecutters.

1912
Middleburg Stone Company opened a new quarry in North Ridgeville, Ohio.

1914
Amherst Cut Stone Company began operations and also specialized in cut stone.

Walter Nord began work for the Cleveland Stone Company as a master mechanic. During his eight-year tenure, he started the American Specialty Company and the General Stone Company and later the Nordson Corporation which has had a rich tradition of corporate and community leadership and wide-ranging philanthropy.

1919
Ohio Quarries Company purchased both the Blum and Delbridge Cut Stone Company and the Amherst Cut Stone Company. The new company was called the **Ohio Cut Stone Company**.

1920
Amherst population was 2,485, Brownhelm Township 974, and South Amherst 944.

1924
Ohio Cut Stone Company opened a quarry in Birmingham, Ohio.

1926
Nicholl Stone Company opened a quarry in Kipton, Ohio.

April 17, 1929
Cleveland Quarries Company was formed, taking over the operations of both the Cleveland Stone Company and the Ohio Cut Stone Company.

1936
Cleveland Quarries Company celebrated its Golden Jubilee (50 years) taking into account the year the Cleveland Stone Company was formed.

1942
Cleveland Quarries Company reduction in cut stone orders resulted in the Plant 4 of The Ohio Cut Stone Division to be leased to a defense material manufacturer for the duration of World War II.

1944
Cleveland Quarries Company reduction in sales was due to a labor shortage caused by World War II rather than a decrease in orders.

1946
Cleveland Quarries Company secured a contract to furnish sandstone for construction of the new John Hancock Mutual Life Insurance Company building, Boston, Massachusetts. This 26 story high building required in excess of 100,000 cubic feet of sandstone.

July 1953
Dr. Jacob O Kamm, nationally known financial economist, joined the Cleveland Quarries Company as a vice president. He was promoted

BEREA *Sandstone*

A NATURAL STONE FOR ADDING BEAUTY AND PERMANENCE TO ALL ARCHITECTURAL PROJECTS

Lone derrick at the No. 6 Quarry in Amherst township in 2005 (Jim Hieb collection).

to executive vice president in October. One of Kamm's daughters was also married to George Steinbrenner, now well known as the owner of the New York Yankees baseball team.

Other 1953 changes included: the company's general offices were moved from Cleveland to South Amherst and a new subsidiary, Silica Chemicals, Inc. was formed. Its initial product is Silicamix, a lining material for foundry cupolas.

1954
Cleveland Quarries Company moved its cut stone department from Amherst to South Amherst.

November 30, 1954
Cleveland Quarries Company sold the net assets and business of its Sterling Abrasives Division.

1961
The Bay of Pigs invasion of Cuba, in an attempt to overthrow Fidel Castro, proved unsuccessful. As the largest importer of grindstones in support of Cuba's sugar cane industry, the Cleveland Quarries Company's grindstone business suffered as a result of the new embargo.

1970
Amherst population was 9,902, Brownhelm Township 826, and South Amherst 2,913.

April 30, 1992
Cleveland Quarries Company shut down operations for a short time.

1995
Kipton Quarries reopened, and a new production facility was constructed in 2006.

1996
Thomas H. Roulston II, purchased 400,000 shares (valued at $1.5 million) for part interest in the Cleveland Quarries Company. He then created **American Stone Corporation** to operate the quarries.

Amherst sandstone was used in the home of Bill Gates, founder of Microsoft.

2002
Amherst sandstone was used in the construction of Oberlin College's new science center on its campus in Oberlin, Ohio.

2003

City of Amherst receives "Sandstone Center of the World" designation from the Ohio Historical Society. An official marker was placed outside of the Amherst City Hall.

2004

British Developer Neil Pike, **Trans European Securities**, announced plans to convert the nearly 1000 acres of rough woods interrupted by five sandstone quarries, into luxury housing, hotels, office buildings, restaurants, and a world-class golf course.

2006

Cleveland Quarries Company donated a large collection of artifacts to the Amherst Historical Society (AHS). AHS solicits the support of Youngstown State University to assist with archival efforts.

The Firelands School District agreed to a tax increment financing arrangement with the county for the proposed quarry project.

2007

January—**Amherst Historical Society** received a grant from the Ohio Historical Society to fund the "Quarry Story," a traveling exhibit compiled from the Cleveland Quarries donation.

January—**Quarrytown.net** website introduced by author of this book.

March—the **Cleveland Quarries Company**, a unit of Amherst Stone Company (AMST), announced the arrangement with Trans European Securities to develop the land owned by the company would not proceed. Local residents speculate if project is really dead.

April—the **Cleveland Quarries Company** announced that it would be moving it's South Amherst operations that includes cutting, storing, and distributing stone from its quarry in Florence Township to Vermilion. This investment includes the purchase of new wire-gang saws which will dramatically increase the output of cut stone. Russ Ciphers, Sr., president and CEO of the company stated, "stone from this region has been used for over 170 years and the demand for stone from the Cleveland Quarries is on the increase. When you

The CLEVELAND QUARRIES COMPANY
1740 East Twelfth Street - - - - Cleveland 14, Ohio

At the end of day, these quarry water boys used buckets, shovels, and other handtools to pretend they were in a band (photo courtesy of Art Koppenhafer).

combine a superior product, a new facility, and a three million dollar investment, you'll be hearing about the success of the Cleveland Quarries for years to come."

The Vermilion location includes a 172,000 square-foot building that resides on 15 acres of land.

The Cleveland Stone Company—One of two thousand bonds is issued by The Cleveland Stone Company and Indiana Quarries Company. This gold bond (No. 1423) was issued on December 1, 1910, and signed by company President, George H. Worthington. The bearer was promised the sum of $1000 in gold coin of the United States of America when it was presented on its maturity date at the offices of The Citizens Savings & Trust Bank in Cleveland, Ohio. A $30 semi-annual interest payment was also provided for 20 years.

The Gold Bond certificate is shown on the next page.

1423

THE CLEVELAND STONE COMPANY,
OF OHIO, AND
INDIANA QUARRIES COMPANY,
OF INDIANA.

$1000

FIRST AND REFUNDING MORTGAGE

PER 6 CENT

GOLD BOND

PRINCIPAL DUE AS WITHIN INDICATED.

INTEREST PAYABLE

JUNE 1st, DECEMBER 1st

PRINCIPAL & INTEREST
PAYABLE AT THE
OFFICE OF
THE CITIZENS SAVINGS & TRUST COMPANY,
CLEVELAND, OHIO.

SUBJECT TO REDEMPTION ON ANY INTEREST
MATURING DATE PRIOR TO MATURITY AT
101% AND ACCRUED INTEREST.

Amherst Historical Society collection.

Penny postcards were common in the late 1800s and early 1900s. This card features the caption "Ohio Quarry, Amherst, Ohio" (courtesy of Fay Ott).

Berea Sandstone trademark adopted by the Cleveland Quarries Company in 1955. This trademark was used to promote the use of Berea Sandstone to members of the architectural community. It stated: "Berea Sandstone adds a quality presentation out of proportion to its cost. Here is permanence combined with attractiveness that means client satisfaction throughout the years. College buildings for example, must carry into future generations the same appeal which Berea Sandstone presents in such an artistic manner. Obtainable in various shades of gray or buff and in patterns available in no other building material. Berea Sandstone is a popular suggestion where the unusual is to be joined with the practical" (Amherst Historical Society collection).

2 SANDSTONE

What Is It—How Is It Used—Where It Was Used

What is it?

There are three general rock or stone categories recognized according to their mode of origin. This is a genetic classification, and it not only states how and under what general conditions a stone was formed, but also implies a general compositional range. The basic stone groups are: igneous, sedimentary, and metamorphic.

Sandstone is a sedimentary rock. It is specifically defined as:

> **Sand-stone** \-ston\ *n* (ca. 1668) : a sedimentary rock consisting of usually quartz sand united by some cement (as silica or calcium carbonate).
> - Webster's Ninth New Collegiate Dictionary, 1984.

In more general terms, sandstone is composed primarily of quartz crystals bonded together with silicon, calcium carbonate or iron oxide. The color of sandstone is determined by the amount of iron oxide present, and the colors can range from light to dark including browns, reds, buffs and yellows. The density, porosity and hardness of sandstone varies from quarry to quarry.

Sedimentary rock can be either of two types: detrital sedimentary stone or chemical sedimentary stone. Sandstone falls into the detrital category which is the naturally cemented accumulation of solid granular materials or particles derived from both mechanical and chemical weathering of any existing rock.

In the dimension stone or cut stone Industry there are many stone types used for applications ranging from countertops, floors, bathrooms, cladding, pavers, fireplace surrounds, and others. The most common stone types include: granite, marble, serpentine, slate, travertine, and quartz-based stones. Sandstone is included in the quartz-based stone group. Other quartz-based stones include: Quartz, Crystalline Quartz, Quartzite, Metaquartzite, Quartz Pebble Conglomerate, Metaconglomerates, Chert, Agate, and Flint. The Quartz-based family of stones are among the longest-lasting minerals in the environment, resisting chemical degradation and abrasion.

How is it used?

During the 1950s, The Cleveland Quarries Company indicated that it sold sandstone for the following uses:
* Bessemer Coverter Linings
* Cut Stone

- Chemical Vat Linings
- Soaking Pit Linings
- Building Stone
- House Ashlar
- Pugh Type Ladle Linings
- Garden Stone
- Cupola Blocks
- Sidewalk
- Silicamix
- Flagging
- Silica Sand
- Laboratory Table Tops
- Breakwater
- Curbing
- Highway Aggregate

Cleveland Quarries Company corporate logo.

In the United States in 1995, dimension stone was produced by 145 companies at 234 quarries in 35 states. Total U.S. output was 1.16 million metric tons valued at $233 million. **Sandstone represented 12.5% of this total output.** The breakdown for the dominant stone types was as follows:

Granite .... 495,000 metric tons
Limestone .... 363,000 metric tons
Sandstone .... 145,000 metric tons
Marble .... 39,600 metric tons
Slate .... 35,600 metric tons

According to the Cleveland Quarries Company, the chemical analysis or make-up of Amherst Sandstone is:

Silica 93.00%
Alumina 4.00%
Ferric Oxide 1.25%
Lime 0.50%
Titania 0.50%
Magnesia 2.14%

Grindstones

"Keep your nose to the grindstone!": a slogan many of us recognize from early childhood. Grindstones played such an important role in our nation's development. Nineteenth-century farmsteads and households utilized sandstone grindstones to maintain the tools necessary to accomplish the tasks needed for daily life. Most grindstones were approximately two inches thick and twenty-four inches in diameter. The stone was mounted on a wood or metal frame.

Grindstones were also used for industrial purposes such as grinding steel. These industrial grindstones were much larger than their farmstead counterparts. Stones as thick as fourteen to eighteen inches and often six to seven feet in diameter were needed. Many industries benefited from the use of grindstones.

The vast majority of grindstones were manufactured in Lorain and Cuyahoga County quarries. But other areas also played a significant role during this time-period: quarries in Cumberland County in Nova Scotia (Canada), Huron County in Michigan, and Washington County, Ohio were all dominate players. Even though man-made grindstones of abrasives such as silicon carbide and aluminum oxide gained popularity in the early 20th century, the sandstone grindstone

Sketch depicts the use of this sandstone grindstone to sharpen saws, axes, and other tools of the day (Amherst Historical Society collection).

industry still exists today, although most agree that the industry died in the 1960s.

The Constitution Stone Company in Marietta held a patent to a process using a steam-powered channeler, or drill, that revolved the stone around a fixed post to fabricate a circular grindstone. Yet, more grindstones were produced in northern Ohio than anywhere else in the United States.

In the 1940s, the Cleveland Quarries sold household grindstones using the name "Berea Grindstones." Its Berea Abrasives division sold several different grindstones. In marketing grindstones to retailers, promotional literature argued:

"Berea Grindstones are best for the merchant to sell and best for customers to buy. Being the result of over fifty years manufacturing experience and continuous, careful planning and testing, a Berea Grindstone naturally has many features not found in ordinary equipment. It provides a lasting, economical sharpening service that brings more business to your store. Point by point comparison with other brands will quickly convince you that the Berea line has real merit. A study of its patented features will provide assurance of its leadership and profit-building possibilities. Stock the best— Berea!"

In April 1941, the Hall Grindstone Company used the channeler to fabricate grindstones at Quarry No. 4 outside of Constitution & Marrietta, Ohio (Amherst Historical Society collection).

Retailers today would have been proud of the retailing techniques encouraged by Berea Abrasives to its retailers. Techniques included:

- Layouts for various floor displays.
- Reminders to allow prospects to try out a Berea Grindstone.
- Sample window displays including instructions on set-up: "Keeping the smaller units in the foreground of the window, and placing the large grindstones in the rear permits a unified presentation that is pleasing to the observer....and profitable to the dealer.
- Designing window displays that attracted men and women.
- Promotions like "Free Sharpening Week—Come In".
- Suggestions to "have an attractive girl, suitably dressed, for the sharpening."

Marketing messages included:

- Berea Grindstones are Best!
- Berea Grindstones are always longer lasting!
- Berea Grindstones are Sturdier!
- Berea Grindstones are Practical and Efficient!
- Berea Grindstones are Smooth Running!

Berea Abrasives marketed a variety of grindstones models including:

- The Motorized Streamliner
- The Harvester
- Monarch Bench Grinder
- The Streamliner
- The Sterling
- The Norka
- The No. 115
- The Bi-Treadle
- The Harvest King
- The Eureka
- The Samson
- The Little Giant
- The Wood Frame Power Grindstone
- The Hercules
- The Keystone
- The Shipstone
- The Family Grindstone

This 1870's drawing depicts a young boy helping his father work the grindstone to sharpen tools for the farm. This advertisement highlighted the role grindstones played in our nation's early history.

Opposite: 1872 Advertisement for grindstones from the Worthington & Sons Company, North Amherst, Ohio (Amherst Historical Society collection).

Advertising Sandstone's Role in Steelmaking

One of the hidden treasures in the Amherst Historical Society's collection from the Cleveland Quarries Company (CQC) archives are advertisements promoting the use of sandstone in the steelmaking industry. Long before the advent of computers and graphic design programs, the hand-sketched ads were as unique as the stone they advocated.

Sandstone played a solid role in steelmaking as it was used for the linings of furnaces, cupolas, converters and ladles used in the mills. The advertisements showcased the use of stone for soaking pit linings. In the late 1950s, CQC supplied three fourths of U.S. steel mills and foundries, as well as mills overseas in countries like Australia and England.

The temperatures in the pits average about 2300 degrees F., so it is remarkable to recognize that the composition of sandstone quarried in Lorain County could withstand the intense heat.

TAILOR-MADE TO FIT THE PIT. HOLDS HEAT BETTER AND LONGER, LESS MAINTENANCE

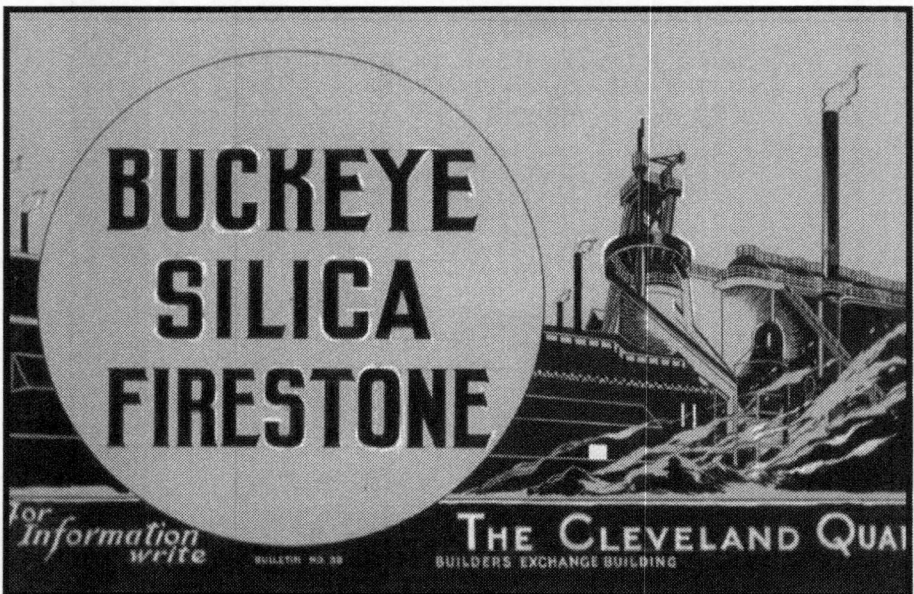

BUCKEYE SILICA FIRESTONE

for Information write

BULLETIN NO. 39

THE CLEVELAND QUAI

BUILDERS EXCHANGE BUILDING

Cleveland Quarries advertisements promoting sandstone in Steel magazines (Amherst Historical Society collection).

Where was it used?

Amherst Sandstone has been used in some of the finest buildings. The following is a sampling of buildings built beyond the borders of Lorain County:

- John Hancock Mutual Life Insurance Building, Boston, MA
- Buffalo City Hall, Buffalo, NY
- Long Beach, California Auditorium (1931)
- The Cleveland Board of Education, Cleveland, OH
- Hockey Hall of Fame, Toronto, Canada
- Senate Entrance at Parliament Hill, Ottawa, Canada
- Cornell University, Cornell, NY
- Detroit Water Works Building, Detroit, MI
- Old City Hall, Toronto, Canada
- Lorain-Carnegie Bridge, Cleveland, OH
- Carolina Life Insurance Building, Columbia, SC
- Wood County Court House, Bowling Green, OH (1895)
- Union Club, Cleveland, OH (1905)
- State Capitol, Lansing, MI (1872)
- City Hall, Buffalo, NY (1930)
- The Bank of Montreal, Waterloo, Ontario, Canada (1914)
- Post Office, Winnipeg, Manitoba, Canada (1905)
- Old Stone Church, Cleveland, OH (1853)
- St. Paul Episcopal Church, Cleveland, OH (1875)
- Sacred Heart Seminary, Detroit, MI (1923-24)
- McKinley High School, Canton, OH (1910)
- Notre Dame College, University Heights, OH (1927)
- University of Michigan Campus Chapel, Ann Arbor, MI
- St. Clare Novitiate School, Rochester, MN (1953)
- Suspension Bridge, Cincinnati, OH / Covington, KY (1865)
- Ohio-Buick Building, Cleveland, OH (1927)
- Brier Hill Steel Company, Youngstown, OH (1923)
- Erie Railroad Station, Jamestown, NY (1931-32)
- Ohio Turnpike Administration Building, Strongsville, OH
- Library and Fine Arts Building—Albion College, Albion, MI (1938)
- Wayne County Court House, Detroit, MI (1897)
- City Hall, Niagara Falls, NY (1923)
- Yale Library

Locally in Lorain County, some of the buildings where Amherst Sandstone has been used include:

Amherst

- Quigley House (1829)*
- Central School (1872)
- Grange Hall (1879)*
- Old Stone Church (1882)
- St. George Chapel (1882)*
- Workshop Playhouse (1882)
- Amherst Town Hall (1884)
- Amherst Library (1906)
- City Hall Bandstand (1915)
- Amherst Hospital (1917)
- WWI Memorial (1924)
- San Spring Building (1924) - originally used as a Catholic School (now City of Amherst Administrative Offices)
- Old Post Office (1939)
- Citizens Home & Savings Bank (1960)

* Building now part of the Amherst Sandstone Museum Center Village

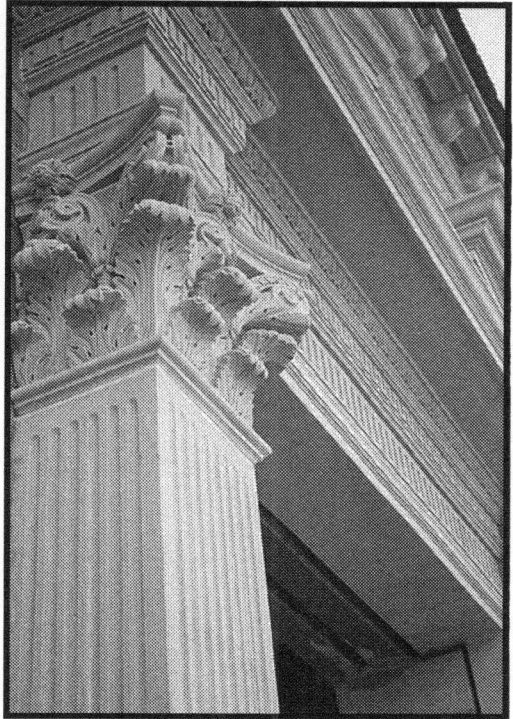

The pages that follow will depict many of the buildings that were adorned with Amherst sandstone over the years. Today, the Cleveland Quarries are still fabricating decorative pieces such as this column showcasing that craftsmanship is still a strong component of the work being done! (photo courtesy of *Stone-World* magazine).

South Amherst
- South Amherst Fire Station
- South Amherst High School
- Old School House
- Makruski's Building (1958)

Oberlin
- Eastwood Elementary School (1956)
- U.S. Post Office (1933)

- Carnegie Library (Oberlin College)
- Finney Chapel (Oberlin College)
- Peters Hall (Oberlin College)
- Oberlin College Science Center (2002)

Elyria
- Lorain County Court House (1870)
- Stoney's Lounge (1956)
- St. Andrews Church

Lorain
- Oak Hills Country Club (1956)
- St. Mary's Academy (1922-23)

Additionally, many of the breakwater walls along the Great Lakes were created using Amherst Sandstone.

Old Stone Church (First Presbyterian Church)
Erected in 1836, the sandstone exterior of the Old Stone Church, adjacent to Public Square in downtown Cleveland, has stood the test of time for 180 years. The sandstone is believed to have been quarried by the Black River Stone Company in Carlisle Township, just south of Elyria, Ohio. The stone has withstood two major fires in the 1800s. The Old Stone Church still stands today, however, you won't see as many 1920 circa automobiles. Too bad this picture does not capture some of the Cleveland Railway Company's streetcars, which also served as a reliable means of transportation for Clevelanders during this era.

Old Stone Church in downtown Cleveland. Amherst, Ohio also has a church known as the "old stone church" (Amherst Historical Society collection).

In May 1931, Contractor E.E. Campbell was busy using Amherst Sandstone to construct the Municipal Auditorium in Long Beach, California. Notice the stone break-wall, as the Auditorium overlooks the Pacific Ocean (Cleveland Quarries Inc. Photograph, Amherst Historical Society collection).

Erected in 1914, Nicholson & Curtis Cut Stone Co. used 6000 cubic feet of Gray "Canyon" Sandstone for The Molsons Bank (now "The Bank of Monreal") in Waterloo, Ontario, Canada (Amherst Historical Society collection).

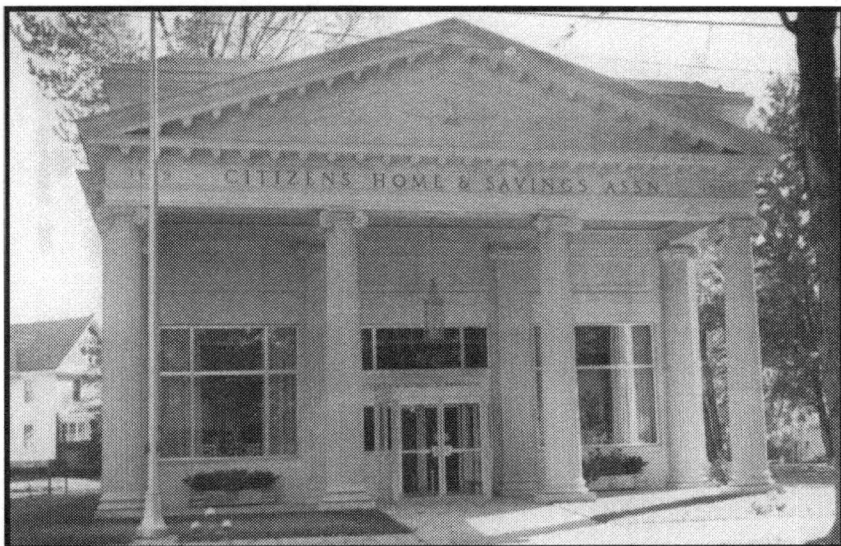

In 1960 Amherst Gray Sandstone was used in the construction of the Citizens Home & Savings Association Bank (now Fifth Third Bank) on Cleveland Street in Amherst (Ludlow Studio Photograph, Amherst Historical Society collection).

A 1927 view of the Motors Realty Building (Ohio-Buick) in Cleveland, Ohio. The historic Buick sign and cars provide almost as much charisma as the Buff Sandstone used on the exterior (Amherst Historical Society collection).

Heyns Bazaar, Detroit, Michigan—Erected in 1861 entirely of Gray Canyon Sandstone that was furnished from the quarries of The Cleveland Stone Company. As this April 1909 picture shows, the interior was entirely burned out down to the ground floor. The Mansard Roof (slate) also collapsed except for the front corner (Cleveland Quarries Inc. Photograph, Amherst Historical Society collection).

Buffalo City Hall was erected in 1930 using "Buckeye Gray" Sandstone produced from The Cleveland Quarries Company and fabricated by The Ohio Cut Stone Company. The building was designed by Architects Dietel & Wade and Sullivan W. Tenes Associates (Amherst Historical Society collection).

Erected in 1924-25, this 1930 photo of the Akron Municipal Building is an excellent display of Amherst Gray Sandstone and Variegated Buff. Sandstone from the Amherst area has been used extensively in government buildings throughout Ohio and beyond. While the list is extensive, in Ohio the list includes municipal buildings and post offices in Avon, Bucyrus, Cleveland, Elyria, Fostoria, Hamilton, Mt. Vernon, Springfield, Portsmouth, N. Randall, Oberlin, Parma Heights, Sandusky, and Troy to name a few. Cities outside Ohio include Lansing (MI), Detroit (MI), Buffalo (NY), Niagara Falls (NY) Fairmont (WV), and Winnipeg (Canada). (Peck Studio Photograph, Amherst Historical Society collection).

This pictures shows how Buckeye Gray Sandstone was often used as decorative art. This carved piece was used on the United States Post Office building in east Akron, Ohio (Undated Photo, Amherst Historical Society collection).

Gray Canyon Sandstone was used for the Wood County Court House, Bowling Green, Ohio built in 1895 (Amherst Historical Society collection.)

"Justice"—Once a block of sandstone—resided over the 3rd Street entrance of the Cuyahoga County Prosecutor's office in the old Court House. One of Cleveland's early landmarks, "Justice" was a mute and convincing testimony of the merits of the stone from which it was carved. Careful scrutiny reveals the sculptor's tool-marks as sharp and prominent today as when the lady of justice made her debut in Cleveland (Ernest Graham Studio Photograph, Amherst Historical Society collection).

Used in more than just government buildings and prestigious college campuses, Amherst sandstone has been used in buildings which include brewing companies and bowling lanes. The Standard Brewing Company (Bottling Department) in Cleveland was constructed by Mogg Cut Stone Company using Amherst Gray Sandstone. Stoney's Rainbow Lanes & Lounge in Elyria has doubled in size since the 1957 picture (Amherst Historical Society collection).

Amherst and Berea Sandstone has been used in many prominent homes, including this town house in our nation's capital, Washington D.C. (T.A. Mullett Photograph, Amherst Historical Society collection).

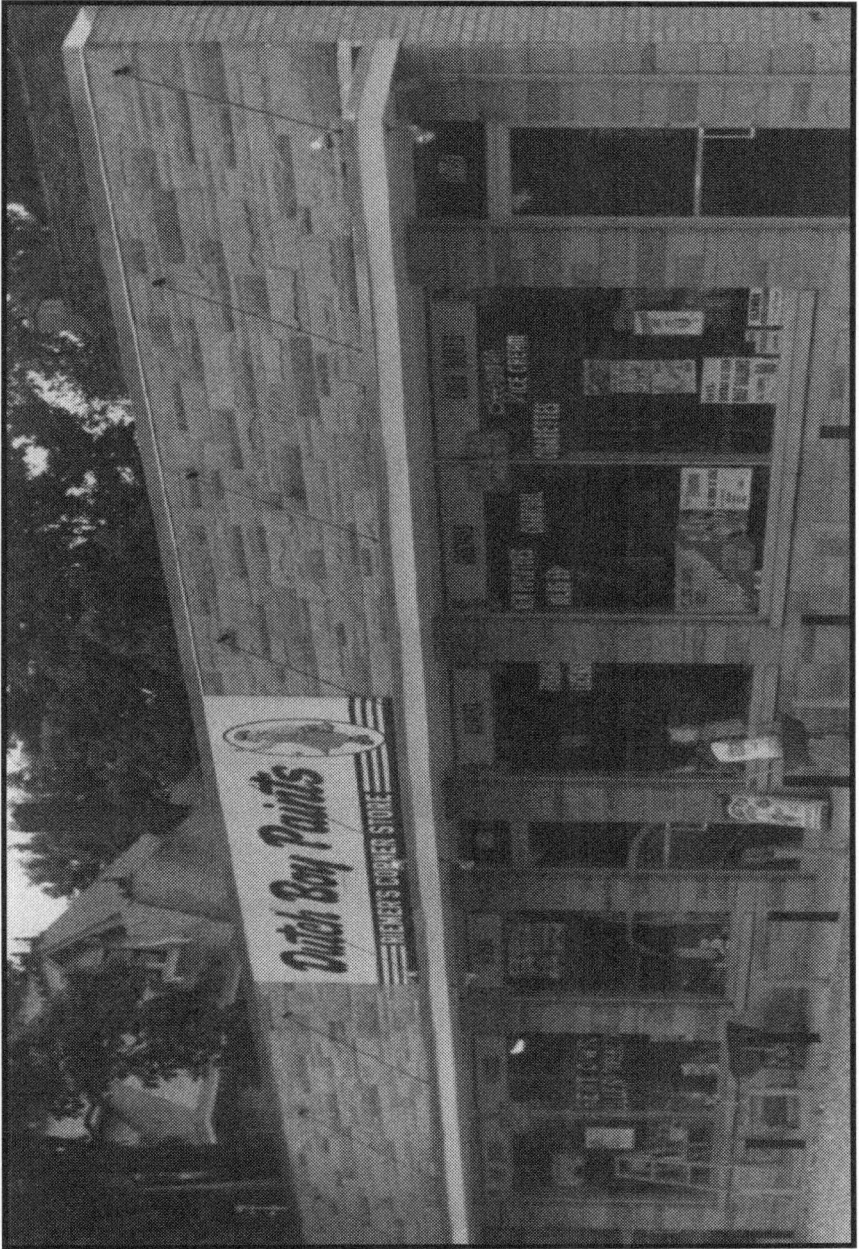

In April 1958 a stone split-face ashlar facing was installed on Reimer's Corner Store in downtown South Amherst. Clearly Dutch Boy Paints were available, as well as fishing licenses, cigarettes, and 10 lbs of potatoes for $0.69 (Ludlow Studio Photograph, Amherst Historical Society collection).

Oberlin Savings Bank—South Amherst branch. This building is now houses a credit union (Ludlow Studio Photograph, Amherst Historical Society collection).

Old Schoolhouse in South Amherst—one of several schools constructed with sandstone in Lorain County in the 1890s (Jim Hieb collection).

Visitors to downtown Amherst may not immediately recognize the former site of the Lorain County Savings & Trust Bank. Constructed with approximately 1560 cubic feet of Birmingham Hard Gray and Variegated Buff Sandstone, this building is now home to the Cedar Pub (Amherst Historical Society collection).

South Amherst High School, now Firelands Middle School (Jim Hieb collection).

CUT STONE DEPARTMENT

Location	Fenton, Michigan	Job No. 2751 (K-3836)	
Name of Project	St. John's R. C. Church		
Kind of Stone	Amherst Buff Sandstone - smooth & rock faced finishes and approx. 436 tons split face Ashlar		
Architect	Smith & Becker		
Contractor	Hornich Construction Co.,-G.C. Jones & Simpson, Mason Contractors		
Estimated Cu. Ft.	Exterior- 2088-0 Interior- 1210-0	Contract Price $	19,376.64 11,011.00
Date Contract	Sanctuary- 689-0 Total Cu.-3977-0 January 5, 1965		6,201.00 36,588.64 Total
	Plus Ashlar -		17,309.00 53,897.64

In 1965, $53,897.64 worth of stone was used for the St. John's Church in Fenton, Michigan. This represented 3977 cubic feet of Amherst Buff Sandstone and 436 tons of split-face ashlar. Churches throughout the region were constructed using sandstone from Lorain County quarries (Amherst Historical Society collection).

Oberlin College's Foundation in Stone

Oberlin College, founded in 1833, was the country's first coeducational college. It was also one of the first colleges to admit African-Americans. Oberlin has a long history of being socially responsible—it can lay claim to many firsts.

Just a few miles south of the famed Amherst stone deposits, Oberlin has many campus buildings constructed with Amherst sandstone. Perhaps a new designation should be made for the college—Built with Amherst Stone!

Oberlin College's Finney Chapel is depicted in this December 1950 picture after a snowfall (Amherst Historical Society collection).

A 2006 view of Finney Chapel on the campus of Oberlin College. The chapel was constructed in 1907-08. Cass Gilbert was the architect of this chapel, as well as many other campus buildings using Amherst Stone (Jim Hieb collection).

On the corner of Lorain and Professor Street, the former Carnegie Library building, now owned by Oberlin College was one of many buildings on the campus of Oberlin College constructed with Amherst sandstone (Jim Hieb collection).

Opposite: In 2002, Oberlin College used Amherst sandstone in the construction of its new Science Center. This project involved the demolition of half of the existing science building and integrating three existing buildings into one science facility. They used the same sandstone that adorns the Cass Gilbert buildings from the early 1900s (Jim Hieb collection).

The Man Who Sold Stone

There was a man who lived by the side of the road and sold stone. He was hard of hearing so he had no radio. He had troubles with his eyes so he read no newspapers. But he sold good stone.

He put up signs on the highway telling "How good The Stone Was." He stood on the side of the road and cried: "Buy Stone." And people bought. He increased his stone orders. He bought more equipment to take care of his trade.

He finally got his son home from college to help him out. But then something happened. His son, said, "Father, haven't you been listening to the radio? Haven't you been reading the newspapers? There's a big depression. The European situation is terrible. The domestic situation is worse."

Whereupon the father thought, "Well, my son's been to college, he reads the papers and listens to the radio, and he ought to know." So the father cut down on his stone orders, took down his advertising signs, and no longer bothered to stand out on the highway to sell his stone. And his stone sales fell almost overnight.

"You're right, son," the father said, "We certainly are in the middle of a great depression."

This story was found amongst the many papers donated by the Cleveland Quarries Company to the Amherst Historical Society. The caption beneath the story read, "I hope your judgements are on what you know and not what you hear?"

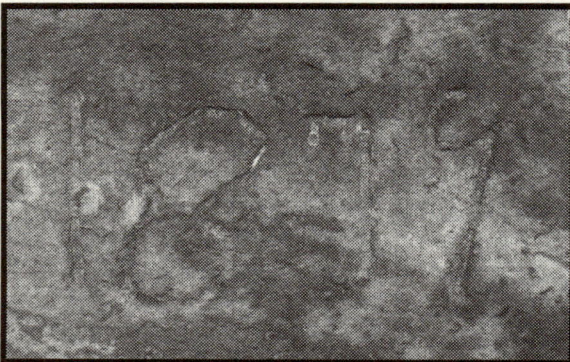

One of the quarry workers carved his name and the year "1879" into the stone of one of the quarry walls at the Brownhelm Quarry (courtesy of Norm & Pat Mulder).

3   LORAIN COUNTY QUARRY SITES

"Amherst was built on a rock"

"The stone quarried...forms such a colossal interest as to have brought together enough people to form a village of 1,500 inhabitants. North Amherst has grown almost entirely from the stimulus given by the development of the stone, and as it is both literally and figuratively built upon a rock, its safety of existence and prosperity is assured. A great change has taken place since the first small openings was made for the purpose of taking out stone...Hundreds of men, assisted by the giant slave, steam, are toiling in the ledges and pits, taking out the rough stone which under the chisel of the artisan shall be conformed into shapes of grace and beauty and strength, to lend majesty to the buildings in the great marts of trade."

History of Lorain County, Ohio, 1879

During its heyday, Buckeye Quarry was believed to be the deepest in the world at 240 feet deep. Today we know that others are much deeper—for example, the E.L. Smith Quarry in Barre, VT, is over 600 feet deep. While Buckeye Quarry is the most famous of quarries in the county, most are unfamiliar with the fact that over 50 quarries were at one time in operation. During the 1870s, many people purchased small tracts of land seeking their fortune. Locals referred to this time period as the "Gold Rush" of the sandstone industry.

Many of the smaller quarries are now hidden in the dense brush, grasses, and trees of the local landscape. Over time, many of the quarry interests exchanged hands through the years until one primary interest controlled virtually all of the quarry land.

Beyond the quality of the stone that was being extracted from the earth, why and how the sandstone industry was able to flourish is largely attributed to the close proximity of Lake Erie. Transporting the stone to other parts of the country and world was possible with the lake being so close. The railroad also played a huge role in the success of the industry.

As research was being conducted to validate the City of Amherst's claim to receiving the State of Ohio's designation "Sandstone Center of the World," a list showcasing current and abandoned quarries in the Amherst region were cataloged. This list was compiled by Sally Cornwell (she also authored the opening section of this book). Lots numbers, quarry land ownership, and years were featured. Some past quarry owners and quarry holes have not been recorded due to incomplete records. As additional research is completed at the Amherst Historical Society, it is anticipated that this list will continue to evolve and grow. We have selected two lots to showcase the level of detail that went into the research project that was critical

LOT 59–60 There were 11 quarries in this area (Ref J) Quarry No. 5 & No. 9 (Ref I & J)
- 1848 John Elliott Quarry (Ref A page 336)
- 1863 William James Quarry (Ref A page 336)
- Parks & E.C. Foster (Ref A page 336)
- 1868 George E. Hall & Company (Ref A page 336; Ref H page 2)
- J. D. Bothwell (Ref A page 338; Ref C)
- 1867 Nichol Quarry (Ref C)
- 1871 Worthington & Sons/Amherst Stone Works (purchased another quarry in this area in 1853) (Ref A page 336 & 337; Ref B; Ref C; Ref E page 176; Ref H page 3)
- 1904 Cleveland Stone Company (Ref B; Ref E page 176)

LOT 93 (three quarry holes) Quarry No. 6 Gray Canyon This quarry is reputed to be the largest quarry in the world. (Ref H page 3) 3,090 feet in length and 1,056 feet wide and 160 feet deep (Ref H page 8)
- Sam Kendeigh (Ref A page 328 & page 338; Ref C)
- Haldeman & Son (Ref A page 328; Ref C)
- 1847 Henry Warner Quarry (Ref A, page 336, Henry Warner bio; Ref G)
- 1869 R.P. Wilson owned another quarry just north of this (Ref H page 3)
- 1874 Clough Stone Company that later becomes part of No. 6 (Ref C)
- 1879 Joseph Barber opens small quarry that later becomes part of No. 6 (Ref A page 338)
- 1886 Cleveland Stone Company (Ref B; Ref D page 86)
- 1903 John R. Walsh; Ohio Stone Company Quarry (Ref B; Ref H page 3)
- 1919 Ohio Cut Stone (Ref H page 4)
- 1929 Cleveland Stone Company (Ref E page 176)
- 1955 Cleveland Quarry Company

REFERENCES:
- A: Book: History of Lorain County, Ohio 1807 – 1879
- B: Map of Amherst dated 1896
- C: Map of Amherst dated 1874
- D: Book: Amherst's Story by Robert Armstrong dated 1914
- E: Book: Lorain County, Her Beautiful Children, Progressive People and Marvelous Development dated 1906
- F: News article: Stone Age dated 1958
- G: News article: Historically Speaking…by historian Bertine Foster
- H: Booklet: The Cleveland Quarries Company dated 1991

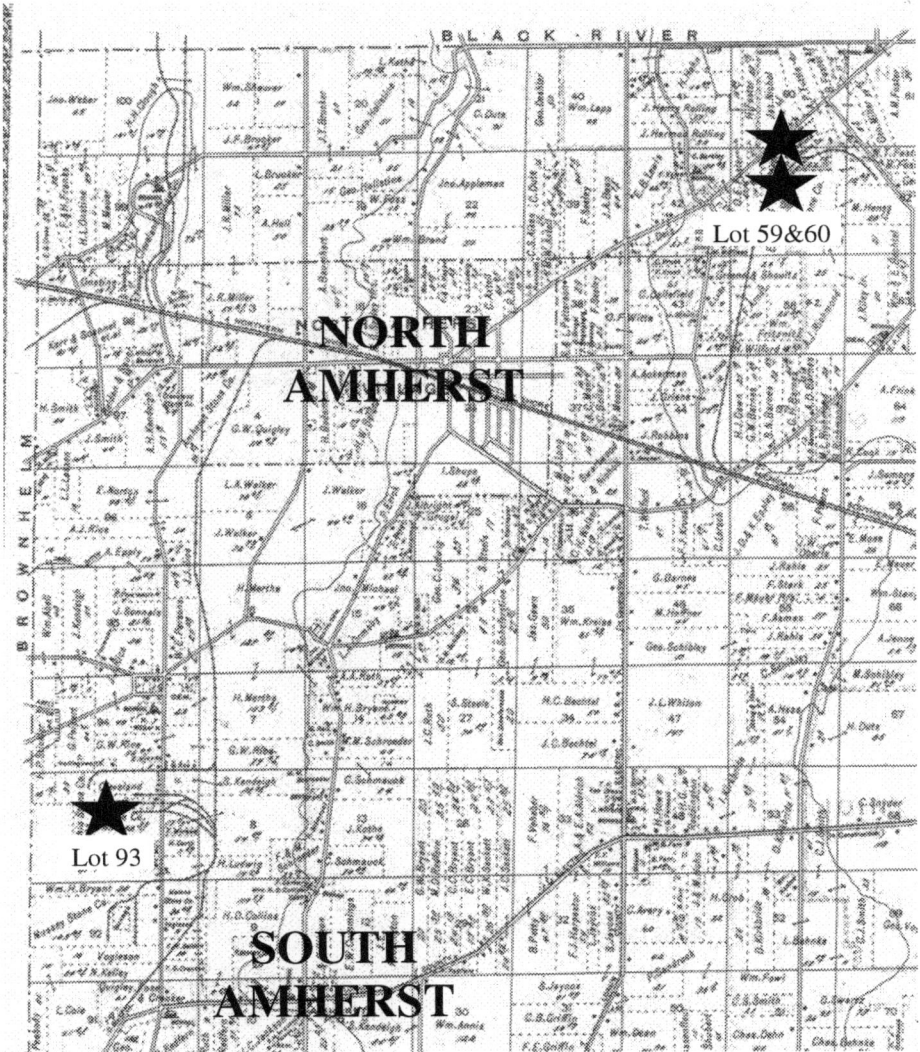

Brownhelm Quarry
No. 10 Quarry

The first quarry in Lorain County was opened in 1847 by Henry Warner in Brownhelm. In 1853 the quarry would be purchased by John Worthington. From this quarry it is believed that the first stone was shipped to a client in Canada. Warner would later become associated with Baxter Clough in his quarry business. He died in 1876.

The Brownhelm Quarry would later become one of many quarries owned and operated by the Cleveland Stone Company and was referred to as the No. 10 Quarry. In 1894, the No. 10 Quarry comprised sixty-two acres, of which approximately eight acres had been quarried to a depth varying from thirty to sixty-five feet.

A 1886 map when the land was owned by the Cleveland Stone Company.

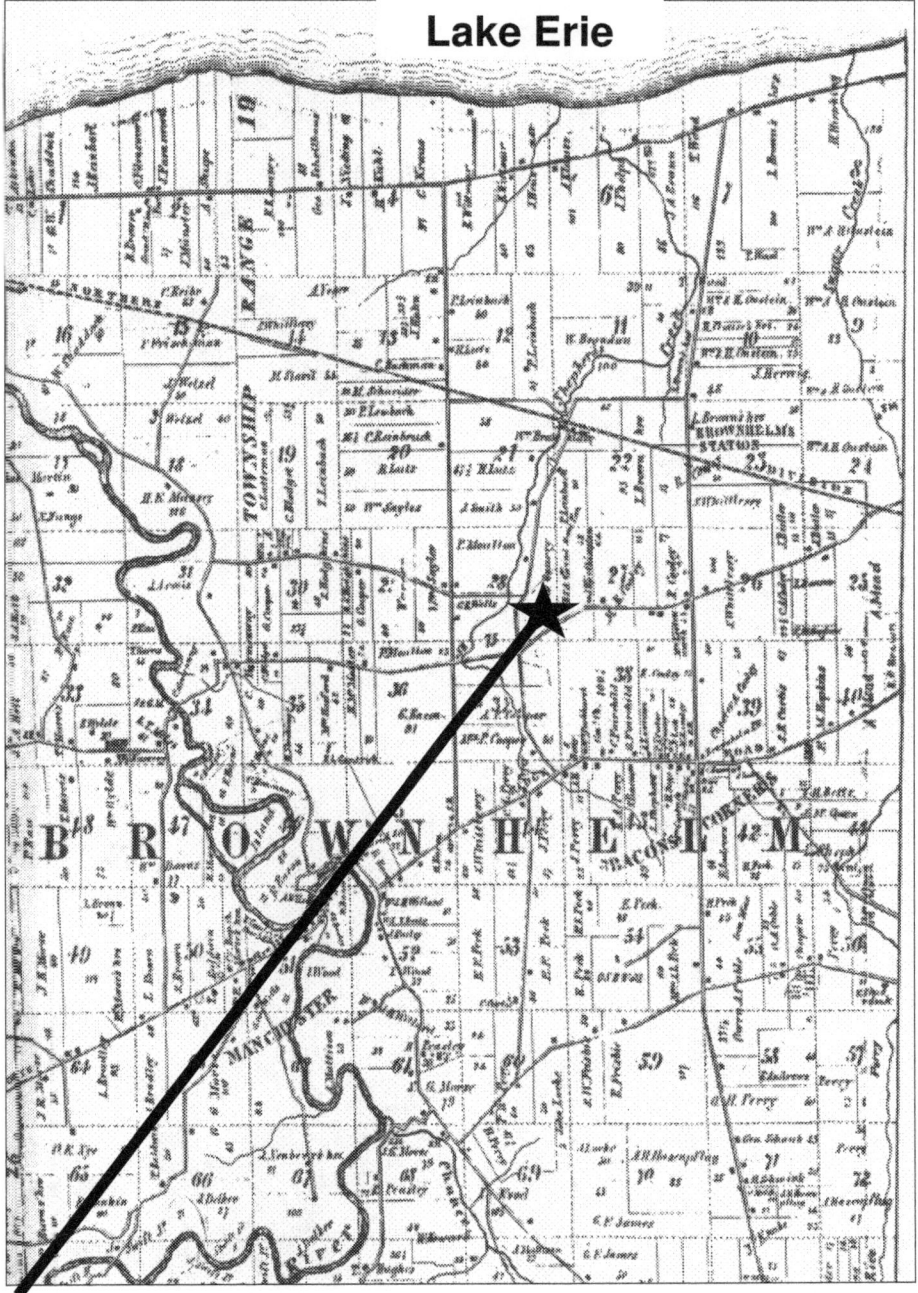

Lake Erie

1857 Brownhelm Township Map depicting the site of the Brownhelm Quarry which later become known as the No. 10 Quarry and part of the Cleveland Stone Company (No. 10 Quarry by Sally Cornwell).

No. 10 Brownhelm Quarry—This undated photograph shows wooden derricks, as well as over twenty men employed at this site. This quarry was the first quarry developed in Lorain County (Quarry #10 by Sally Cornwell, photo courtesy of the Brownhelm Historical Society).

Hidden amongst the remains of the Brownhelm Quarry is this stone bucket which was used by quarry workers to hoist scrap stone from the quarry bottom. One of the sandstone deposits in this area had a blue tone that was extremely popular. Unfortunately, the quarry workers hit a natural water spring and eventually this portion of the quarry had to be abandoned because all pumping efforts could not keep up with the spring water (Jim Hieb collection).

Chapter 3 - Quarry Sites in Lorain County   **77**

Worthington Quarries

The Worthington family would play a major role in the quarrying operations in Lorain County. After purchasing the Brownhelm Quarry, they also obtained interests in Huron County, Michigan, and other quarries in Lorain County. George Worthington was instrumental in the incorporation of the Cleveland Stone Company, a merger of several Lorain County quarry companies.

Several maps donated by the Cleveland Quarries identified quarries simply by the name "Worthington Quarry" indicating the scope of the Worthington interests (such as the one below).

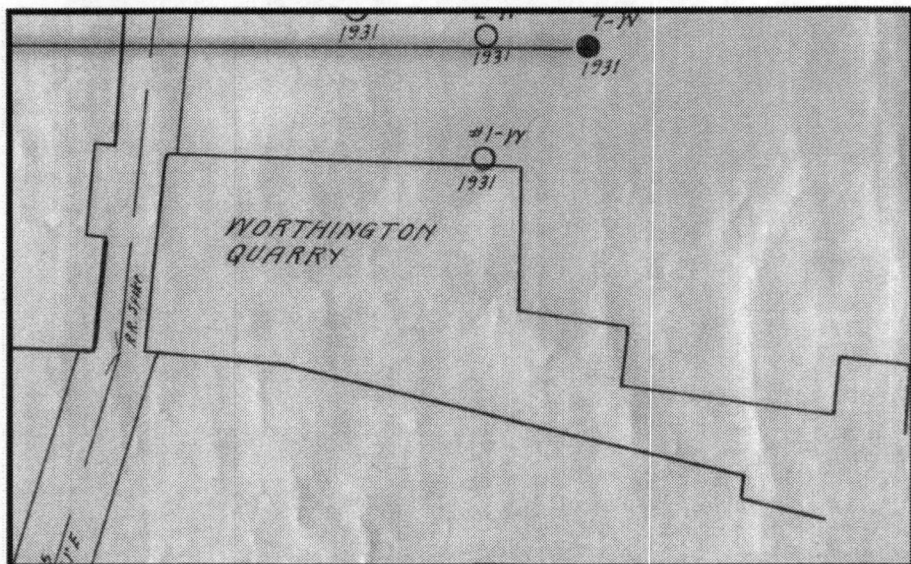

This Worthington Quarry is off of Quarry Road just north of the machine shop (Amherst Historical Society collection).

Grindstone Sold in Tons
In the early 1870s, the Worthington Quarries were producing:
- 1871—7,950 tons of grindstones
- 1872—9,762 tons of grindstones
- 1873—9,260 tons of grindstones

During the same time period, their block sales were 188,233; 247,239; and 205,490 cubic feet respectively.

Clough Quarries
...and the Pony Trail

Baxter Clough was not a native of the Amherst area, but was influential in the local marketplace. A native of New Hampshire, he and his family moved to Ohio in 1839 and later to Berea in 1946. He had an inventive mind claiming a stove, called the "Regulator," as one of his inventions. A failed business venture manufacturing pig iron led him to the manufacturing of small grindstones. This introduction to the stone industry led him to the quarries at North Amherst in 1852.

Three quarries would be associated with the Clough name in the Amherst region:
- 1852—initial quarry opened; this quarry would later become known as the Clough Middle Quarry.
- 1869—two additional quarries opened; these quarries would be known as the Clough North and Clough South Quarries.

The North and Middle quarries were located in North Amherst (near Oak Point Road), while the South Quarry was located in South Amherst.

Clough's creative nature led to the invention of machinery and applied steam for turning large grindstones, as well as other types of machinery to improve the quarry operations. In 1860, the Clough Quarry was generating 3000 tons of grindstones. Initially the grindstones were hauled by mule teams to Black River (now Lorain) and shipped to Cleveland. Clough was one of the first quarry owners to build a dock at the lake so that the stone could be transported by water.

The grade from the Clough Quarry to the dock was sufficient that the flat bed cars would go under their own weight, aided by the mules. Demand for block stone for building purposes increased dramatically over the next few years (Clough was not the only benefactor of this demand). The slowness of the mules would soon be replaced by a small train engine.

A spur from the Brownhelm Quarry also had sufficient grade for the flatbed cars to travel to the water docks under the weight of the stone. The cars would return to the quarry with the aid of horses or mules. The reference "Pony Trail" can be found in many historic documents.

The Clough Stone Company owned its own boat that sailed for many years, but a storm claimed the boat and its stone contents.

Clough also purchased and developed what become known as the Independence and Columbia Quarries before his death in 1872.

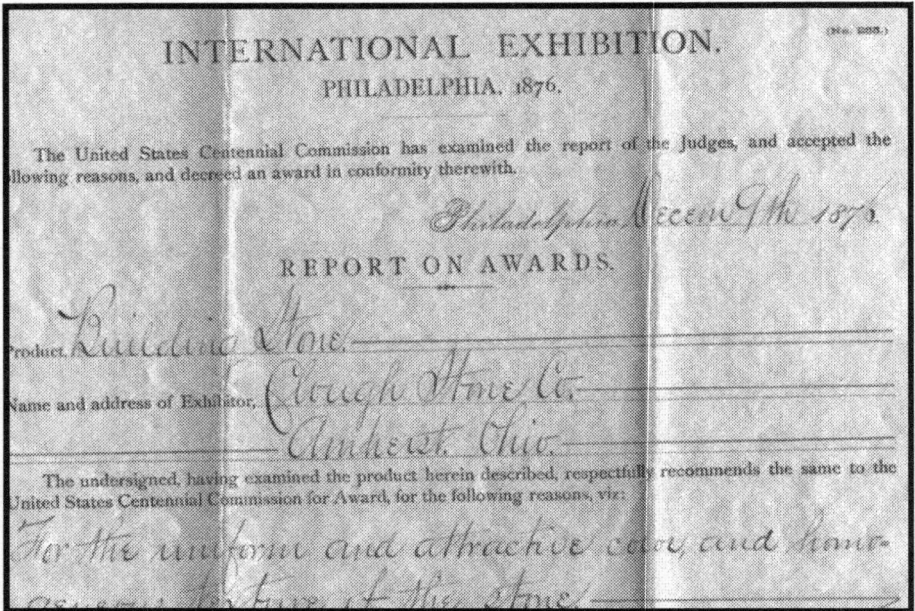

INTERNATIONAL EXHIBITION.

PHILADELPHIA. 1876.

The United States Centennial Commission has examined the report of the Judges, and accepted the following reasons, and decreed an award in conformity therewith.

Philadelphia December 9th 1876

REPORT ON AWARDS.

Product. *Building Stone.*

Name and address of Exhibitor, *Clough Stone Co. Amherst. Ohio.*

The undersigned, having examined the product herein described, respectfully recommends the same to the United States Centennial Commission for Award, for the following reasons, viz:

For the uniform and attractive color, and homogeneous texture of the stone.

Clough Stone Company wins an award at the International Exhibition (Amherst Historical Society collection). Following the end of the Civil War, Americans began to prepare for the celebration of the nation's 100th birthday in 1876. In 1871 Congress enacted a bill forming a commission to "prepare and superintend the execution of a plan for holding an exhibition, and, after conference with the authorities of the city of Philadelphia, to fix upon a suitable site within the corporate limits of the said city where the exhibition shall be held." President Ulysses S. Grant appointed the commission members. The International Exhibition was open in 1876 from May 10th to November 10th. In the category of building stone, the Clough Stone Company was recognized "for the uniform and attractive color, and homogenous texture of the stone."

Opposite—The top picture is a photo of the train that hauled sandstone from the Clough quarries to Lake Erie to be loaded on ships for transport to Canada. Notice that timber was used for this portion of the rail which is very different from the railroad track leading the to No. 6x Quarry depicted approximately 75 years later (Amherst Historical Society collection).

No. 6 Gray Canyon Quarry
No. 6x Quarry

Opened in the spring of 1869, the No. 6 Gray Canyon Quarry (formerly known as Haldeman Quarry) has produced over 455 million cubic feet of sandstone. Adjacent to this quarry was the No. 6x Quarry, as well as a smaller quarry operation known as the Malone Quarry. Over time, as sandstone was removed from these various sites, one massive hole resulted.

The No. 6 Mill which sat atop the quarry to process the sandstone blocks into smaller slabs and other finished products was reputed to have the largest steam engine in the world. This engine provided the necessary power to operate the gang saws.

The winter swimming hole – Quarry No. 6x
Growing up near the quarries, hundreds of kids over the years found innovative ways to keep themselves occupied. Near any quarry operation, ponds are a common by-product and swimming was an activity that surprisingly occurred year-round. It is true that not all of their activities were endorsed by parents.

Inside of the No. 6 Mill in the 1950s. Giant blocks of sandstone that were hosted up from the bottom of the No. 6 Gray Canyon Quarry and cut by gang saws (far right) into slabs (Amherst Historical Society Collection.)

Circa 1910 view of the Malone Quarry, one of the many quarries that now make up the No. 6 Quarry viewed from Quarry Road (Amherst Historical Society Collection.)

Harold Dalton, son of Hurstle Dalton (see Foyster Matlock interview), recounted several stories dating back to the late 1950s and early 1960s about swimming at the quarries in the middle of winter. In his circle of friends, it was common for twelve or so friends to enjoy swimming, coupled with the importance of a nearby campfire. It would start with one-half of the swimming party taking large rocks to break a hole through the ice. Careful attention would be spent to clearing any loose chunks of ice in the water. If not cleared the loose ice could cause serious head injuries from diving. Ice would also be stacked to serve as makeshift diving boards. The remaining members of the party would collect firewood and start a fire on shore. The fire was crucial to tackle the effects of swimming in cold water during the dead of winter.

To the casual observer, one would wonder about hyperthermia and its affects on anyone who remained in freezing water for any length of time. As Dalton commented, "we couldn't spell it [hyperthermia], nor knew what it was. We just focused on watching after one another. If someone's lips turned blue and their teeth chattered, then we made sure they took a turn by the fire." He added that the activity of swimming in the winter was not something he was proud of, nor something he would ever want his children or grandchildren to participate in.

If there was ever the potential for an Olympic swimmer to hail from South Amherst, then one of them could have originated from this group of winter quarry swimmers. Dalton indicated that if another opening existed at the other end of the pond, one of his friends would swim under the ice and emerge at the other end. This type of training is not found in any swimmers training manual! Yet it was a task successfully completed in the winter swimming holes in South Amherst.

Partial view of the No. 6x Quarry (Ernest Graham Studio, Amherst Historical Society Collection.)

Work crews utilize a channel cutter at the No. 6 Quarry in 1957. Cuts as deep as 15 feet are made in the stone by a channel cutter driving a steel ram into the rock as it moves inch by inch along the rails. Notice the lack of safety hard hats and other safety precautions so prevalent today.

Below—sawmill at the Malone Quarry, circa 1910 (Amherst Historical Society collection).

Nichols Quarry

James Nichol, a native of Dundee, Scotland, was a stone cutter who moved to the U.S. in 1852. From 1861 to 1870 he served a management role at the Brownhelm Quarry. In 1867 he had purchased a quarry from Joseph Barber that was adjacent to his farm. In the 1870s Nichols operated his own quarrying venture and became a prominent businessman producing grindstones and exporting sandstone.

The initial fortune made by Nichol did not come from the sandstone industry. He had invested $4,000 in the Beeman Chewing Gum Company. Three generations of Nichols would be involved in quarrying operations in Lorain County. The family also established a foundation to fund charitable endeavors.

Workers believed to be posing outside the Nichols Quarry. Few smiles can be seen on their faces on what must have been a very hot day. Notice the two water boys in the lower left hand corner (Amherst Historical Society collection).

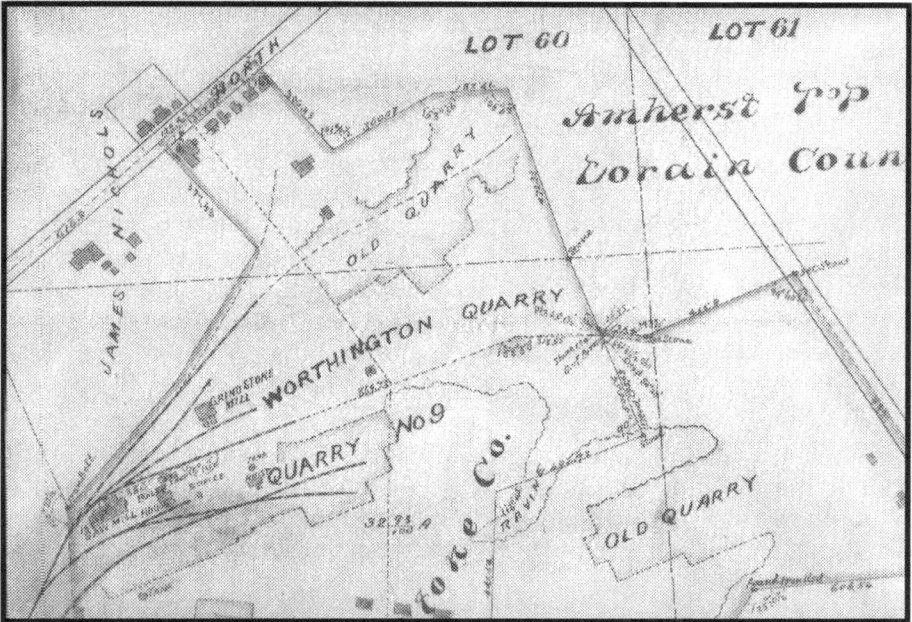

Undated map believed to originate from the 1890s shows the close proximity between the Worthington Quarry and the Nichols Quarry. State Route 2 cut directly between these two quarries. Passersby along the highway will recognize the American flag painted on the south side of the highway (Amherst Historical Society collection).

Quarry ledge of a small abandoned quarry near the Nichols Quarry. Notice the metal handrail that still remains from an old ladder used by workers to reach the quarry bottom (Jim Hieb collection).

Chapter 3 - Quarry Sites in Lorain County   **87**

Buckeye Quarry

An early brochure of the Cleveland Quarries Company best summarized this quarry.

"It is 240 feet deep, 600 feet wide and 1,800 feet long. When one sees the dwarfed size of the trees growing on the quarry bottom, the toy-like appearances of the machines that are used in the operations and the ladders reaching far into the depths, it seems almost impossible to the visitors and tourists who make daily stops at this quarry, that men, in so short a time, could have created such an unbelievable 'hole in the ground' as exists at South Amherst. No place on earth provides a similar panorama to that presented by Buckeye Quarry which is visited by thousands of tourists every year. No other place more dramatically portrays man's victory over Nature in more definite terms than this deep, canyon-like, man-made quarry reaching into the earth. It was started long ago and man's persistence has made it what it is today. During working hours, the men can be plainly seen in the depths, appearing very small because of the fact that this work appears to be dangerous, it is a fact, that because of experience, these workers have not suffered a serious accident in many, many years—a tribute to their skill and the precautions with which they are surrounded. The men working in these depths have the satisfaction that comes from specialized craftsmanship and are justly proud of their jobs."

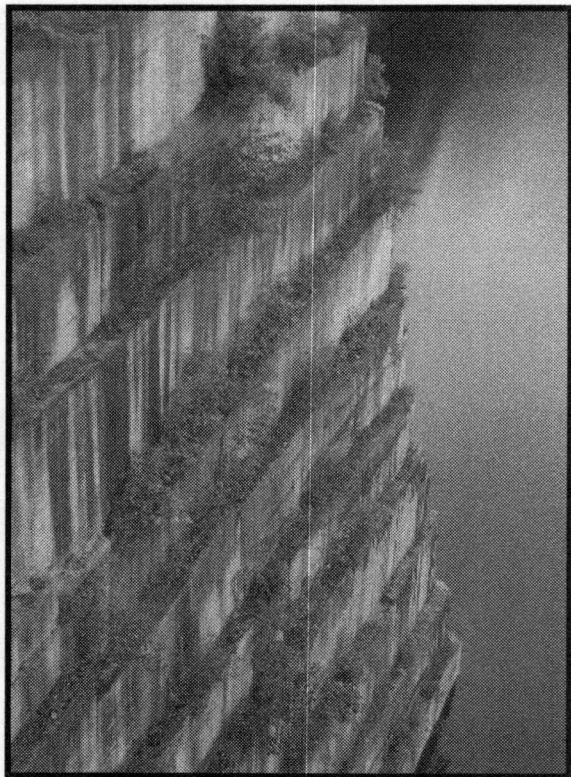

Long abandoned, the Buckeye Quarry ledges are overgrown with brush and the quarry bottom filled with water (2006 picture, Jim Hieb collection).

1909 picture caption denotes the Buckeye Quarry as the "largest stone quarry in the world" (Amherst Historical Society collection).

No. 7 and 7x Quarries

The No. 7 and adjacent No. 7x quarries are situated on property to the north of the famous Buckeye Quarry and south of the No. 6 and No. 6x quarries. These quarries were all sources for Amherst Grey Sandstone. The origin for these quarries began in 1893 when the Mussey Stone Company of Elyria incorporated and began quarry operations in the Amherst district. The Mussey operations would be absorbed by the Cleveland Stone Company four years later.

A close look at these two pictures show both a close-up and more distant view of work being done at the Mussey Quarry, circa 1910 (Amherst Historical Society collection).

View of the No. 7 Quarry in July 1959. Several steel derricks can be seen, as well as the safety railing used by the derrick tender. Bill Sowders served as the derrick tender on this day (see insert) and is posed to use hand signals to direct the derrick operator so that sandstone block can be carefully hoisted from the quarry bottom to the surface above. Housed in a building (see bottom right picture) without direct view of the quarry bottom, the derrick operator manages hand levers relying on the derrick tender to be his eyes and ears for the successful trip from the quarry bottom to the surface (all three photos courtesy of Mary Louise Back).

The $7 Bicycle at Quarry No. 7x

In the 1950s seven dollars was a large amount of money for any kid to acquire. One local youth had scraped up enough money to purchase a used bicycle. Imagine the pride and the bragging rights a new bicycle (even if used) would garner within the neighborhood. You have heard many a story about priceless baseball cards being ruined to turn a bicycle into an imaginary motorcycle. Long before any of us had heard of Evil Kenvil, world-renowned stuntman, there was the $7 bicycle at Quarry No. 7x and the thrill of adventure beyond the affect any Mickey Mantle baseball card could deliver.

When it comes to quarries, you can count on steep cliffs and the development of ponds at the quarry floor. When it comes to kids, you can count on the opportunity for a dare. At Quarry No. 7x, both facts met face-to-face. The dare was to drive the newly acquired bicycle at top speed over the cliff of Quarry No. 7x into the quarry pond some 70 feet below. With an additional water depth of 50 feet, this was a dare New Deal Owens (yes, this South Amherst youth was named for the President Franklin D. Roosevelt's social program of the same name) and his friend Terry Hoover were willing to accept. More than pride and bragging rights were at stake. Somehow several neighborhood boys had collected $5 to wager on this dare.

Assured he would not lose his prized bicycle, New Deal, with Terry riding on the handlebars, rode off the top of the quarry and went splashing into the water. They could not hold onto the bicycle and it was lost forever to the quarry floor.

Grafton Quarry

Stone was also quarried outside of Grafton, Ohio. These pictures were taken around 1910 when the quarry and grindstone mill was owned by the Ohio Quarries Company (Amherst Historical Society collection).

Opposite—2002 view of a portion of the No. 7 Quarry being worked (courtesy of StoneWorld Magazine).

Kipton Quarry

Circa 1910 picture of the Kipton Quarry when the quarry was owned by the Ohio Quarries Company.
Below: The Kipton Quarry scabbing yard as stone blocks are being loaded onto a railroad car. You won't see workers wearing hardhats or other safety equipment during these days (Amherst Historical Society collection).

Mori Quarry
Built to divert the Ohio Turnpike?
Rated the Best Swimming Hole

At the northern most point of the Cleveland Quarries property the Mori Quarry resides just a stone's throw away from the Ohio Turnpike. Travelers along the turnpike probably don't even notice that the road bends in an arc shape because of this quarry. The origin (or legend) behind the quarry is much different than any other in Lorain County.

This quarry is named after P.A. Mori, a long-time administrator with the Cleveland Quarries Company. Dug around 1950, the quarry itself is small—only 30 feet at its deepest point. In what can best be described as "which came first, the quarry or the turnpike," the true origin of the quarry was not in fact to extract sandstone. Rather as development plans were being formulated by the State of Ohio and the Ohio Turnpike Commission to construct Interstate 90 across Ohio, the Cleveland Quarry officials noted that the tentative plans included a portion of the new road being constructed on its property. While this property was not being quarried at the time, the Cleveland Quarries wanted to preserve the acreage for future use and offer some persuasion for the Ohio Turnpike to move its project slightly northward.

A derrick was hastily installed at the site of the proposed road and the Mori Quarry was dug. Although only a couple of men occasionally worked the quarry, it gave a sufficient appearance of a working quarry and was large enough to divert the new turnpike.

The Mori Quarry was also reputed to be the best swimming hole for kids growing up in the region during the 1950s and 60s. Unlike the other well-established, deeper quarries, the shallowness of the Mori Quarry meant the water temperature would be much warmer. The frequent absence of quarry personnel also meant easier access to sneak onto the property for a refreshing swim. Additionally, the quarry was safer than others because the distance from the top of the quarry to the water surface level was also only two ledges.

Mussey Stone Quarry
Elyria

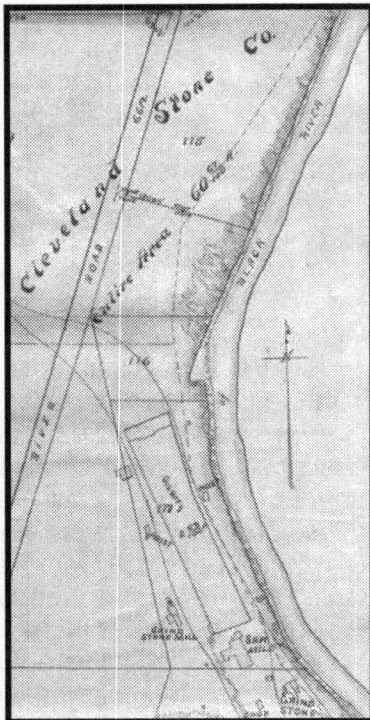

The biggest quarry in Elyria was founded by Henry E. Mussey. It was on the west side of the west branch of the Black River. A swinging foot bridge once existed to provide workers quicker access near West 8th Street. It was later used as a dump by a steel company.

In 1893 Mussey would incorporate the Mussey Stone Company and open a quarry a short distance from the Gray "Canyon" Quarry. The map depicts the Elyria Quarry shortly after the Cleveland Stone Company acquired the assets of Mussey stone operations. The Capital Stock certificate #1 (bottom) was assigned to founder Henry E. Mussey (dated October 2, 1894).

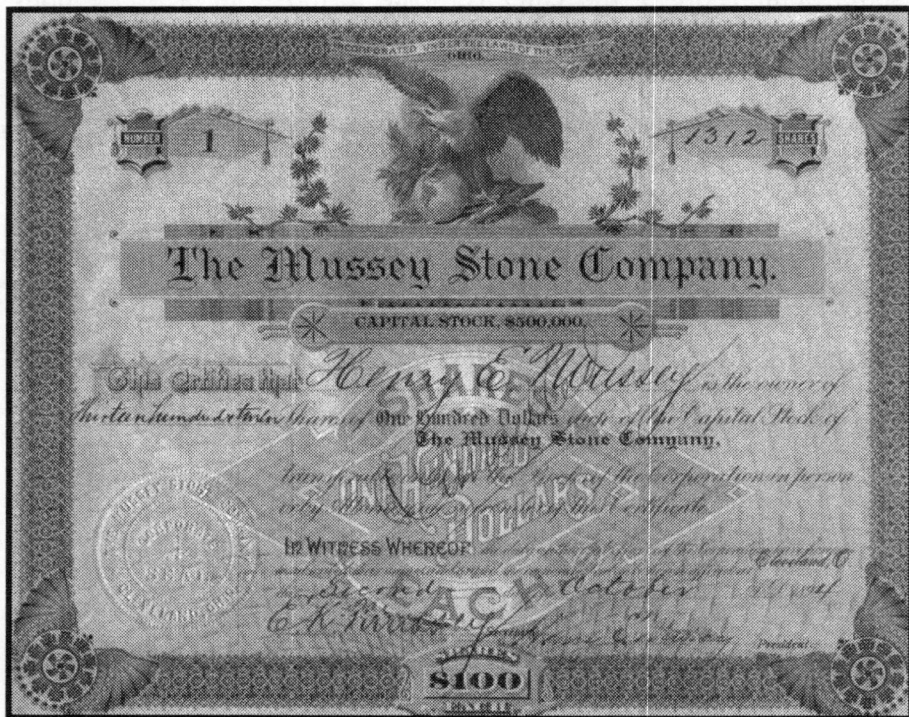

Nickle Plate Quarry
LaGrange Township

In the mid-1870s a small community called Nickle Plate sprang up in LaGrange Township. It was named by James Nichols who was superintendent of the Atlantic Company who operated the quarry company. Ten quarry houses and boarding houses existed for the quarry workers and their families.

The Nickle Plate Quarry was known for its Blue Rock sandstone. The stone and its grindstones were shipped to all parts of the world. The stone was also used for foundations and sidewalks throughout LaGrange village.

In 1884, the Toledo-Norwalk-Cleveland Railroad constructed a railroad spur from Oberlin to the quarry.

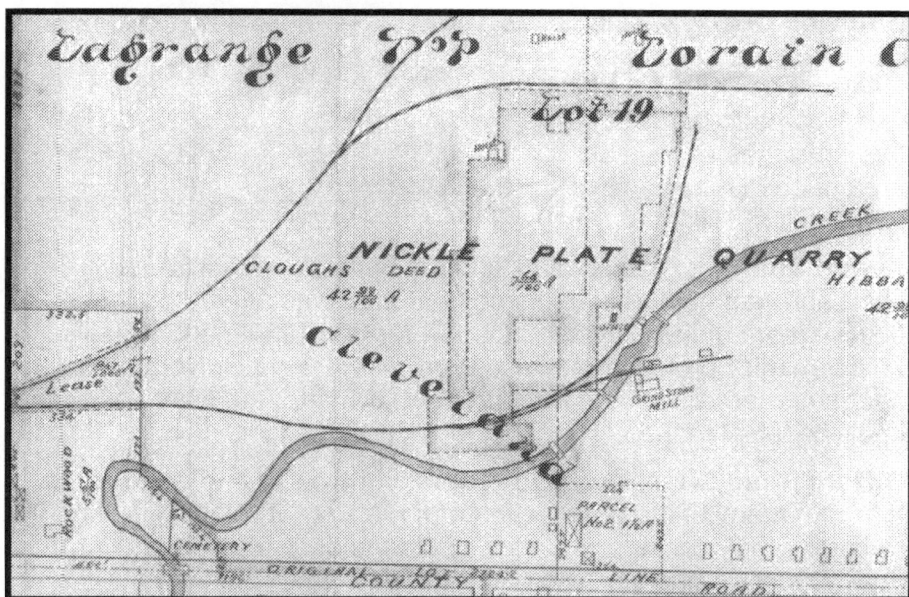

1886 map of the Nickle Plate Quarry near Nickle Plate Road and Whitehead Road in LaGrange Township. It is interesting to note the reference to the "Cloughs Deed" suggesting that Baxter Clough likely played an earlier role in this quarry site (Amherst Historical Society collection).

Other Quarry Notes
Lorain County

W.O. Quarry, was the first stone quarry in Elyria. It was located between Gates Avenue and Ninth Street and its stone was used in building the railroad bridge at East Bridge Street. The quarry has since been filled.

Bryant Quarry was also located in Elyria.

Beaver Creek Quarry was listed on several 1890 quarry maps adjacent southeast of the Buckeye Quarry in South Amherst. It is likely that this small quarry was later absorbed into the famed Buckeye Quarry.

Bothwell Quarry was located west of the Nichols Quarry. The **Ohio Quarry** was located to the south. All three quarries were in an area at one time known as "Green Horn Hill."

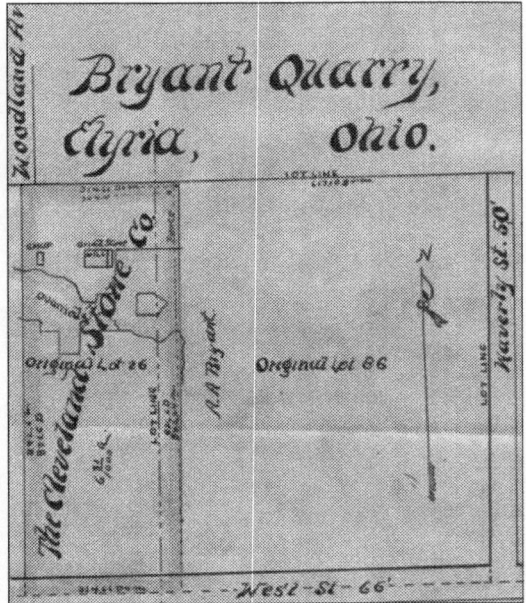

Kamm Quarry is located directly north of the No. 8 Mil in South Amherst and was named after Jacob O. Kamm who served on the executive team of the Cleveland Quarries in the 1950s, and was also an economics professor at Case Western Reserve University. His daughter married George Steinbrenner, owner of the New York Yankee's baseball team.

Gifford Road Quarry was located west of Oberlin and north of Kipton. Like many of the quarries in Lorain County, it was a popular swimming hole for kids who snuck onto the quarry property after the quarry workers had gone home for the day.

4 WHAT YOU CAN SEE FROM THE ROAD

This chapter showcases what can still be seen as your pass by on foot or motor vehicle in 2007. With the Cleveland Quarries plans to relocate their operations to Vermilion, it is likely some of what you can see from the road today will change. We wanted to capture these images and others—see how many of these sites you recognize.

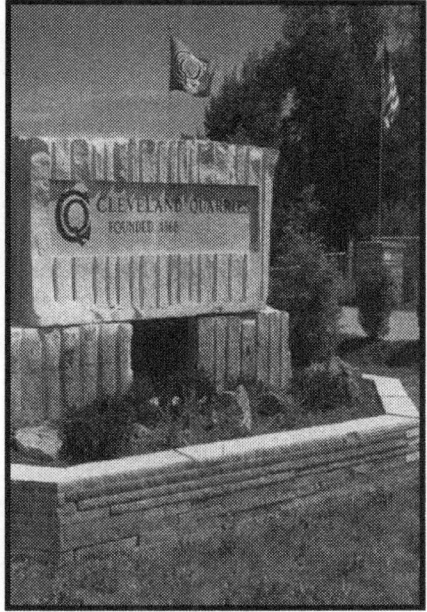

Cleveland Quarries Company Entrance off of State Route 113, South Amherst, Ohio.

Kipton Quarry Entrance off of State Route 511 just north of Kipton, Ohio. The Kipton Quarry re-opened in 1995 after sitting idle for many years. This operation is independent of the Cleveland Quarries.

Take a close look at this circa 1910 picture of the buildings that surrounded the No. #6 Quarry. Most of the buildings have long been torn down, however, the machine shop (below), No. 6 mill (opposite), and the bridge can still be seen from Quarry Road (Amherst Historical Society collection).

2006 picture of the machine shop along the eastern side of Quarry Road. This century old building was built in 1906 (Jim Hieb collection).

The No. 6 mill along the western side of Quarry Road. You can faintly see a guide wire from the lone derrick that remains.

Single derrick at the Mori Quarry can be see from the Ohio Turnpike, just 1/2 mile from Baumhart Road. (2007 photos, Jim Hieb collection).

2006 view of the No. 8 Mill—also viewable from Quarry Road. Approximately 100 yards north of the mill is the Kamm Quarry. Adjacent to the south is another abandoned quarry. This mill is still being operated by Cleveland Quarry Company.

Below—sandstone blocks staged outside the No. 8 Mill (Jim Hieb collection).

View of the stone yard and offices used by the Cleveland Quarry Company.

As you enter the Cleveland Quarries entrance off of State Route 113, visitors approach a fork in the road. This stone marker provides direction to both the No. 8 and No. 3 mills (Jim Hieb collection).

Travelers on State Route 2 just east of Amherst witness the American Flag painted by Barbara (Brucker) and Grant Thompson in 1976 to celebrate our nation's bicentennial. While the flag has occasionally also been painted with graffiti, good samaritans have always touched up the flag following these acts of ill-respect. Following the 911 attacks, Barbara and Grant, along with family members again gave the flag a fresh coat of paint as a testimony of patriotism.

A home near Parsons and Nickle Plate Diagonal Road acknowledges a small abandoned quarry within view of their home (Jim Hieb collection).

5 QUARRYING PROCESS

Early quarrying techniques in Lorain County involved laborers removing the stone by hand. Several stone quarries still display the chisel markings left from these early days. This was especially true of the many outcroppings that were by local farmers to obtain stone for their building foundations and other building purposes.

The commercial quarrying operations at quarries such as Buckeye, Clough, and Gray Canyon used four processes: clearing, blasting, channeling, and fabricating.

Clearing

The sandstone deposits in Lorain County are believed to be between 250 to 400 million years old. The stone deposits were left in giant boulders but were not connected. As a result, the various Lorain County quarries are scattered and unconnected.

When a new quarry was being prepared to be opened, the area would be core-drilled to determine the quality of stone. If the area was determined to have sufficient depth for quarrying, the surface dirt and other materials would be removed. It would not be uncommon for several feet of dirt to be removed. In some cases, explosives would be used to assist with loosening up the stone cap.

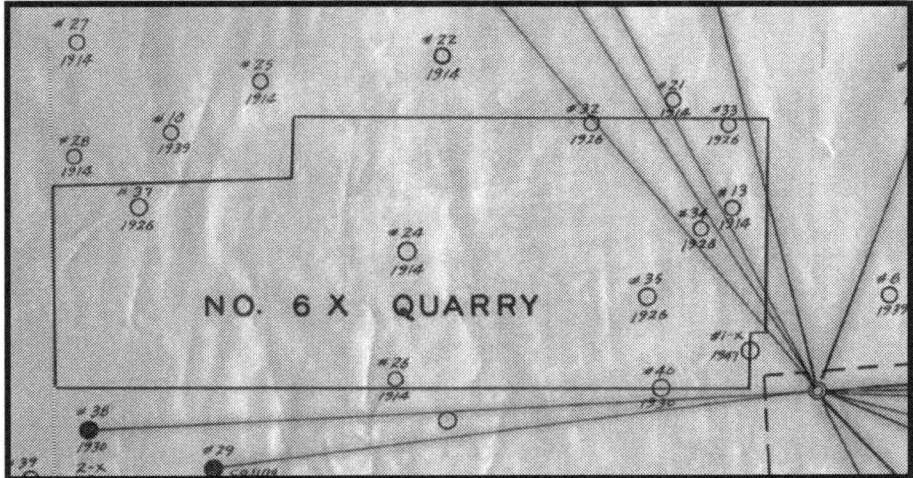

1951 map showing a number of core-drill notations at the No. 6x Quarry in Amherst Township. The earliest notations are from 1914 (Amherst Historical Society collection).

In 1919 the Erie Shovel & Ball Engine Company was used to remove the topsoil at a new sandstone quarry site. Horses and wagons were used to transport the soil and other debris from the area (Amherst Historical Society collection).

Blasting & Channeling

Quarry workers would use a large steam or air-powered chisel that was mounted on rail wheels. This "channel cutter" drove a steel ram into the stone, sometimes as deep as fifteen feet. A crane would then remove the channel cutter from the track so that explosives could be used to widen the cut made by the channel cutter.

The channel cutter would be placed on the tracks again as a parallel channel would be cut about six feet from the previous cut. Another blast with explosives would split the stone block loose. The block would range in length from 12 to 20 foot sections and weigh 12 to 40 tons.

In 1903, the Ohio Quarries Company utilized several channelers manufactured by the Sullivan Machinery Company. Their 1903 catalogue promoting their products stated:

"It is claimed for this machine that it will cut more feet of channel per day than any single or double-gang channeler now made, and with less cost in labor, steam consumption and repairs. Only two men are required to run it, instead of three men as required for double-gang channelers. In very soft stone, however, a third man may be used to good advantage. It is readily moved about or operated in small space, and the head or standard can be set at either end of the machine, to cut into any corner, making the machine either right or left-hand. Owing to these conditions, in connection with its light weight, it is well adapted for opening new quarries, where a larger and heavier machine would not find sufficient room to operate.....The stone channeling machine, used for years in quarrying dimension stone, came into prominence before the engineering public during the construction of the Chicago

Drainage Canal, and has since been employed extensively on work of a similar nature throughout the world."

Derricks, using cables, would lift the blocks and place them on railroad cars to be transported to the fabrication mills for finishing.

Bill Back directs the loading of a sandstone block onto a flatbed car (circa 1957, Mary Louise Back photo).

1903 picture shows two quarry workers posing on the railroad car carrying three sandstone blocks (Amherst Historical Society collection).

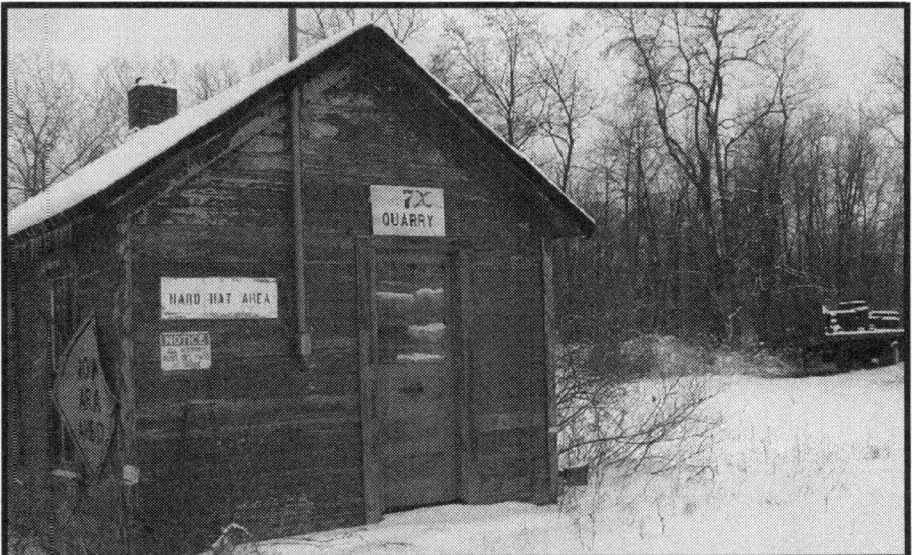

This building was constructed to store the dynamite for the No. 7x Quarry (2005 photo courtesy of Sally Cornwell).

1910 derrick at one of the Amherst quarries (Amherst Historical Society collection).

Hoisting gears located at the northwest corner of the No. 6x Quarry. August 8, 1952 picture (Amherst Historical Society collection).

Fabricating

The stone was cut into different shapes and thicknesses by giant gang saws with steel blades. Early processes included the use of sand and water. Modern gang saws use diamond-studded blades and water.

These saws were used to transform the giant blocks into slabs as the picture below depicts. Smaller sizes could be fabricated by bridge saws. Columns would typically be put on planing machines and lathes, while decorative work would be done by the stonecutters using pneumatic chisels and other hand tools.

Lorain & Southern Railroad served a number of quarries in Lorain County. This circa 1900 picture shows the flatbed railroad cars loaded with sandstone (Amherst Historical Society collection).

Gangsaws used to cut stone blocks into slabs (date unknown—Amherst Historical Society collection).

This picture was used as the cover of the 1951 Cleveland Quarries Company annual report. It depicts workers outside the No. 3 Mill removing stone from a conveyer belt after being cut in the mill (Amherst Historical Society collection).

Quarry Operations—A Small City

As the various quarry companies were consolidated into larger operations, one result was the formation of buildings that resembled a small city. In April 1929, the Cleveland Quarries Company was formed as a merger of the Ohio Quarries Company and the Ohio Cut Stone Company (both the result of previous mergers). An appraisal of the Ohio Quarries Company facilities in South Amherst prior to the merger showed a valuation of:

	Buildings	Equipment	TOTAL
Cost of Reproduction:	$297,486	$641,165	$938,651
Sound Valuation:	$219,050	$432,302	$642,353
Insurable Valuation:	$143,197	$372,260	$515,457

The map on the adjoining page shows the small city that was the Ohio Quarries Company. The buildings included (numbers represent the label on the map):

1—Powerplant and Machine Shop
2—Amherst Mill Building
3—Buckeye Mill Building
4—Grindstone Building
5—Blacksmith Shop
6—Locomotive Shop
7—Curbstone Planing Mill
8—Welding Shop
9—Bar Shed
10—Millwright & Storage Building
11—Crushed Stone House
12—Office & Electric Shop
13—Time Office Building
14—Carpenter Shop
15—Saw & Motor Shed
16—Lime House
17—Supply Shed
18—Pump House
19—Hoist House
20 & 23—Yard Office

21 & 22—Thaw Shed
24—Railroad Yard Office
25, 29, 30—Garage
26—Tool House
27—Railroad Track Scale House
28—Wagon Scale House
31—Excelsior Storage House

THE OHIO QUARRIES COMPANY
UNION TRUST BUILDING   CLEVELAND, OHIO

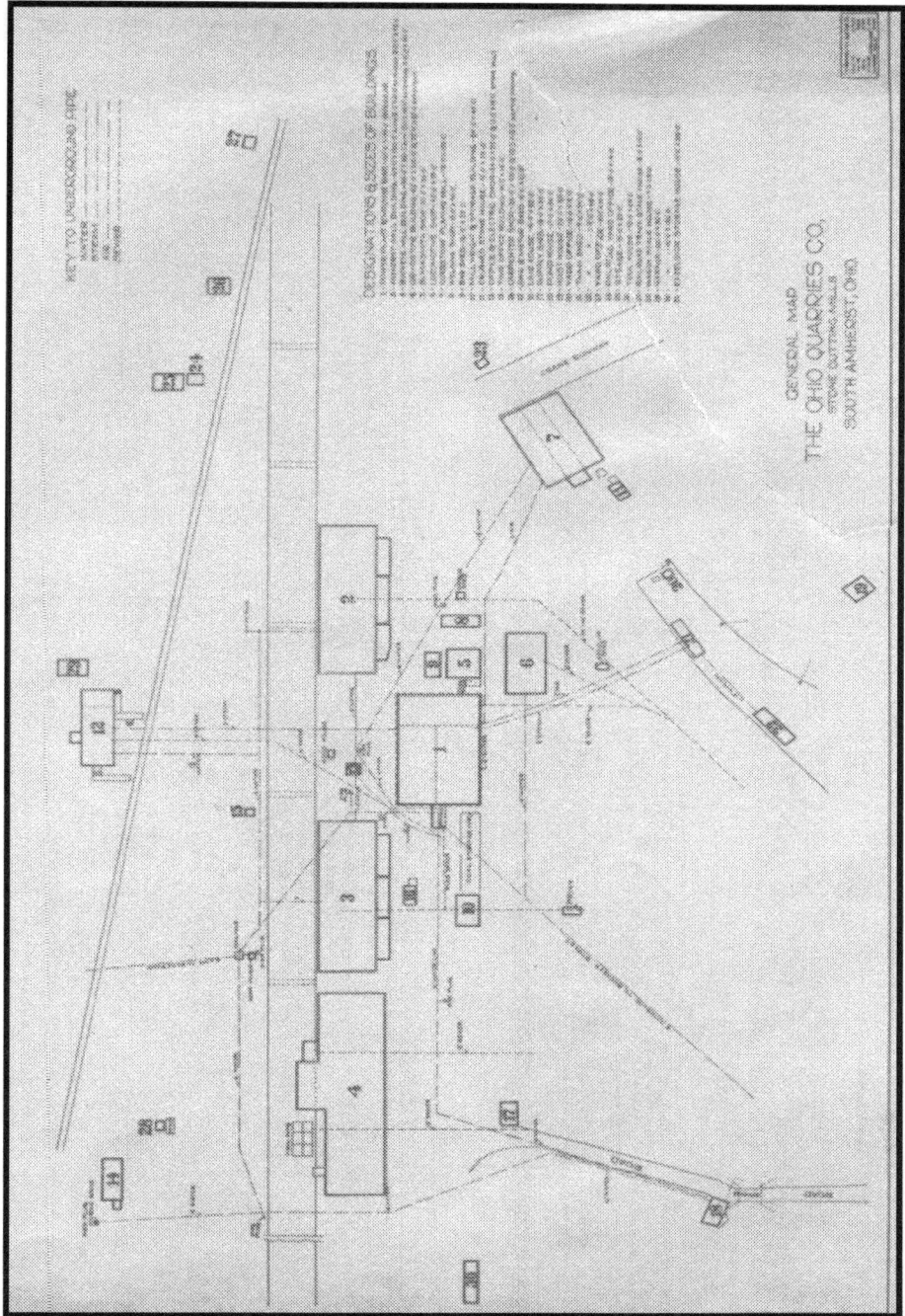

General map showing the Ohio Quarries Company buildings from March 23, 1929, appraisal by The Manufacturers' Appraisal Company (Amherst Historical Society collection).

Baby Drill Operator in 1919 for the Cleveland Stone Company (Amherst Historical Society collection).

SOURCES

Adams, Jennifer, **Entering a third century of quarrying**, Stone-World Magazine, September 24, 2002.

Baldwin-Wallace College, **www.bw.edu**, Berea, Ohio: Baldwin-Wallace College, 2005.

Berea Abrasives, **Berea Abrasives Catalog**, Cleveland, Ohio, 1947.

Berea Historical Society, **website**, Berea, Ohio, 2005.

Bible, King James Version, Indianapolis, Indiana: Kirkbride Bible Co., 1988.

Buswell, Dorothy McKee, **LaGrange Township History**, LaGrange, Ohio, 2006.

Cornwell, Sally, **Quarry No. 10**, Amherst, Ohio: Amherst Historical Society, 2005.

Dakota Granite Company, 2005, Millbank, SD.

Goulder, Grace, **Rocks to Riches in Northern Ohio**, Cleveland, Ohio: Cleveland Plain Dealer Pictorial Magazine, June 20, 1954..

Kropko, M.R., **Business Rebuilt Stone by Stone**, Cleveland, Ohio: The Cleveland Plain Dealer, September 24, 2002.

Manufacturer and Builder, **The Berea and Amherst Sandstone**, Manufacturer and Builder, Volume 19, Issue 5, May 1887.

Marble Institute of America, **Dimension Stone Design Manual, VII**, "**The Geology of Stone**," Westlake, Ohio: Marble Institute of America, 2007.

McGeachy, Liz, **Cleveland Quarries-History, Legacy & Future**, The Slippery Rock Gazette, October 2004.

Morain, Grace, **Amherst Sandstone Quarries**, Elyria, Ohio, Report presented at a Lorain County Historical Society Meeting, October 12, 1958.

Parker, Alex M., **American Stone Plans to Relocate to Vermilion**, Lorain, Ohio, The Morning Journal, April 14, 2007.

Pickrell, Martha M., **Investing in Community: The History and Legacy of the Nord Family of Ohio**, Oberlin, Ohio: Oberlin Heritage Center, 2004.

Sego, Mickey, **Then There Was None**, A history of the Berea Sandstone Quarries, Berea, Ohio: Berea Historical Society, 1996.

Smith, Ron T., **Improving Upon An Old Legacy**, Steelways, August 1957.

Sullivan Machinery Company, **Quarrying Sandstone at Amherst, Ohio**, Mines and Quarry, August 1906.

Sullivan Machinery Company, **The excavation of rock by** machinery, Sullivan Machinery Company Catalogue no. 51, 1903.

The Cleveland Quarries Company, **The Cleveland Quarries Company**, South Amherst, Ohio, 1955.

The Cleveland Quarries Company, **Cleveland Quarries Newsletter**, South Amherst, Ohio: The Cleveland Quarries Company, Fall 2003.

The Cleveland Quarries Company and Subsidiaries, **Annual Reports**, South Amherst, Ohio, 1939-1947 and 1953-1954.

The Grindstone, March/April 2006.

U.S. Department of the Interior (U.S. Geological Survey), **Mineral Industry Surveys**, January 23, 1997.

And....**Amblin Around**, by Connie Davis (a regular column featured in the Amherst Times).

Many reminiscent stories told by friends and family about quarry life in Lorain County.

Hoisting machine built by the Bedford Foundry & Machine Co, Bedford, Indiana (2005 picture courtesy of Sally Cornwell).

ACKNOWLEDGEMENTS

This book would not have been possible without the contributions, advice, and support made by: Howard Akin, Amherst Historical Society, Mary Aufdenkampe, Mary Louise Back, Marcie Becker, Tom Bray, Russ Ciphers Jr., Russ Ciphers Sr., Cleveland Quarries Company, Sally Cornwell, Dakota Granite Company, Harold Dalton, John Dietrich, Garis Distelhorst, Christa Hieb, Kali Hodgson, Pat Holsworth, Art Koppenhafer, Marble Institute of America, Foyster Matlock, J. Russell Matlock, Stan Matlock, Homer E. Morrow, Chuck Monson, Chuck Muehlbaurer, Norm & Pat Mulder, G.K. Naquin, John & Lisa Novak, Oberlin Heritage Center, Fay Ott, Judy Riggle, Mary Roberts, Donna Rumpler, Dr. Claude Rust, Kyle Schwochow, Stone Magazine, and StoneWorld Magazine.

Cover design by Sue Myers

Front cover picture courtesy of Susy Youngless

Back cover picture taken by Jim Hieb

Circa 1919 Wagon Drill used in Berea, Ohio quarries (Amherst Historical Society collection).

ABOUT THE AUTHOR

A native of Tripp, South Dakota, James Hieb received a Bachelor of Science Degree from the University of South Dakota (Vermillion, South Dakota) and a Masters Degree from Bowling Green State University (Bowling Green, Ohio). He moved to Lorain County in 1991 and now serves as Special Projects Director for the Marble Institute of America, the leading trade association serving the natural stone industry. He is an adjunct faculty member at Lorain County Community College and member of the Oak Grove Missionary Baptist Church, Rotary Club of Oberlin, Stonington Lodge, Cub Scout Pack 435, Oberlin Heritage Center, and Amherst Historical Society. Hieb resides in South Amherst with his wife Christa and two sons, Kyle and Chad.

The author didn't always do his research alone. On this trip the Novak girls and Hieb boys were travel companions exploring the quarry countryside in Lorain County. Sydney, Chad, Kyle, and Samantha take a break to lean against an outcropping of sandstone and pose for this picture (Jim Hieb collection).

ONLY A FRACTION OF THE QUARRY STORY IS
ENCLOSED IN THIS BOOK. FOR MORE
INFORMATION ABOUT QUARRY LIFE
IN LORAIN COUNTY,
WE ENCOURAGE YOU TO VISIT:

AMHERST HISTORICAL SOCIETY &
SANDSTONE MUSEUM CENTER
163 MILAN AVENUE
AMHERST, OHIO
WWW.AMHERSTHIST.ORG

OR VISIT OUR WEBSITE AT
WWW.QUARRYTOWN.NET

LORAIN COUNTY

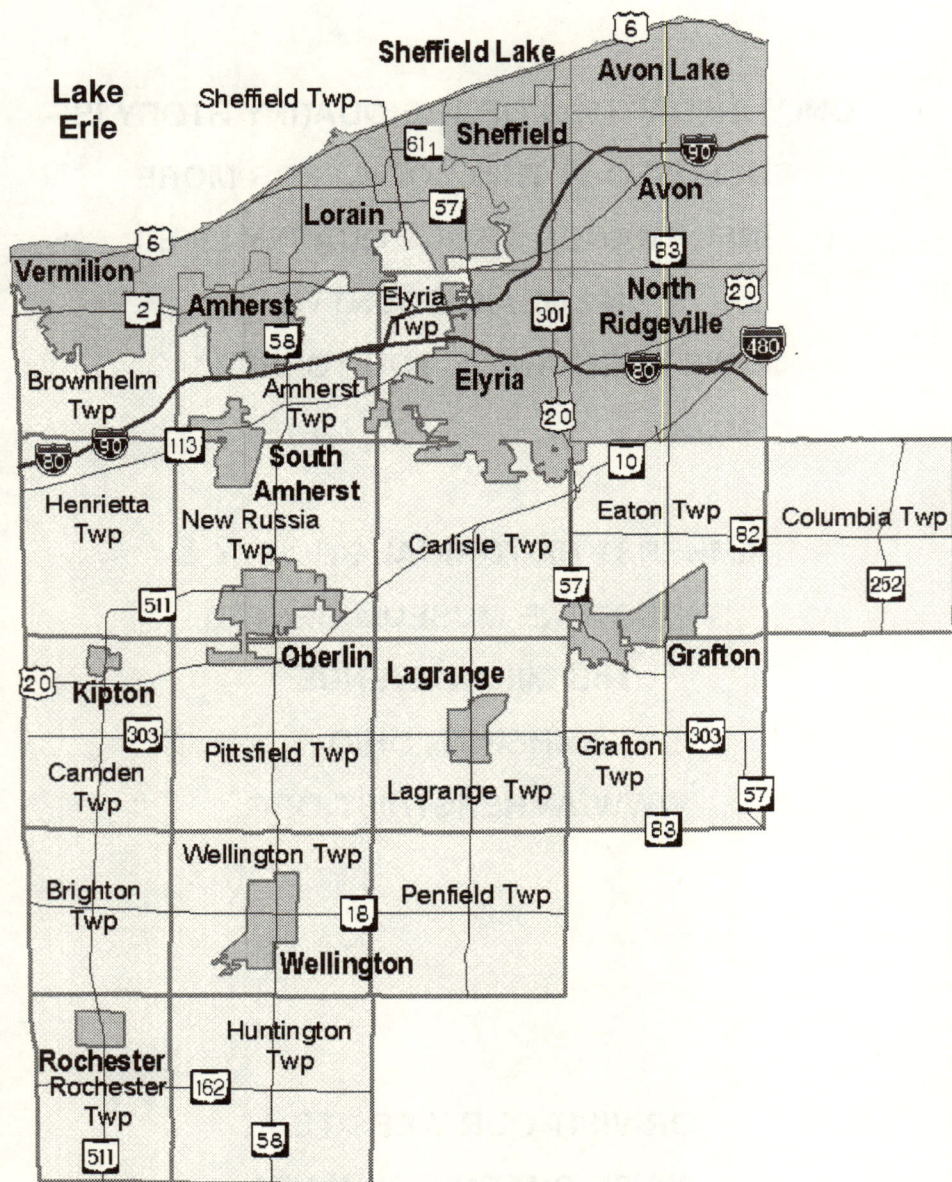

Lake Erie

Sheffield Lake

Sheffield Twp

Avon Lake

6

Lake Erie

61 1

Sheffield

90

Sheffield

57

Avon

Lorain

83

6

Vermilion

Elyria Twp

North Ridgeville

20

2

Amherst

301

58

480

Brownhelm Twp

Amherst Twp

Elyria

80

South Amherst

20

80 90

113

Henrietta Twp

New Russia Twp

10

Eaton Twp

Columbia Twp

511

Carlisle Twp

82

252

57

Oberlin

Lagrange

Grafton

20

Kipton

303

Pittsfield Twp

Grafton Twp

303

Camden Twp

Lagrange Twp

57

83

Wellington Twp

Brighton Twp

18

Penfield Twp

Wellington

Rochester

Huntington Twp

Rochester Twp

162

511

58

CLEVELAND QUARRIES™
YESTERDAY - TODAY - TOMORROW

CLEVELAND QUARRIES A UNIT OF AMERICAN STONE'S PRIMARY BUSINESS IS IN THE QUARRYING AND FABRICATING OF SANDSTONE.

CLEVELAND QUARRIES' LEGENDARY BEREA SANDSTONE™ IS KNOWN FOR ITS EXCEPTIONAL DURABILITY AND STRENGTH FOR USE IN RENOVATION, RESTORATION, ARCHITECTURAL, AND LANDSCAPE PROJECTS.

WE ENCOURAGE YOU TO VISIT:

WWW.AMST.COM

WWW.CLEVELANDQUARRIES.COM

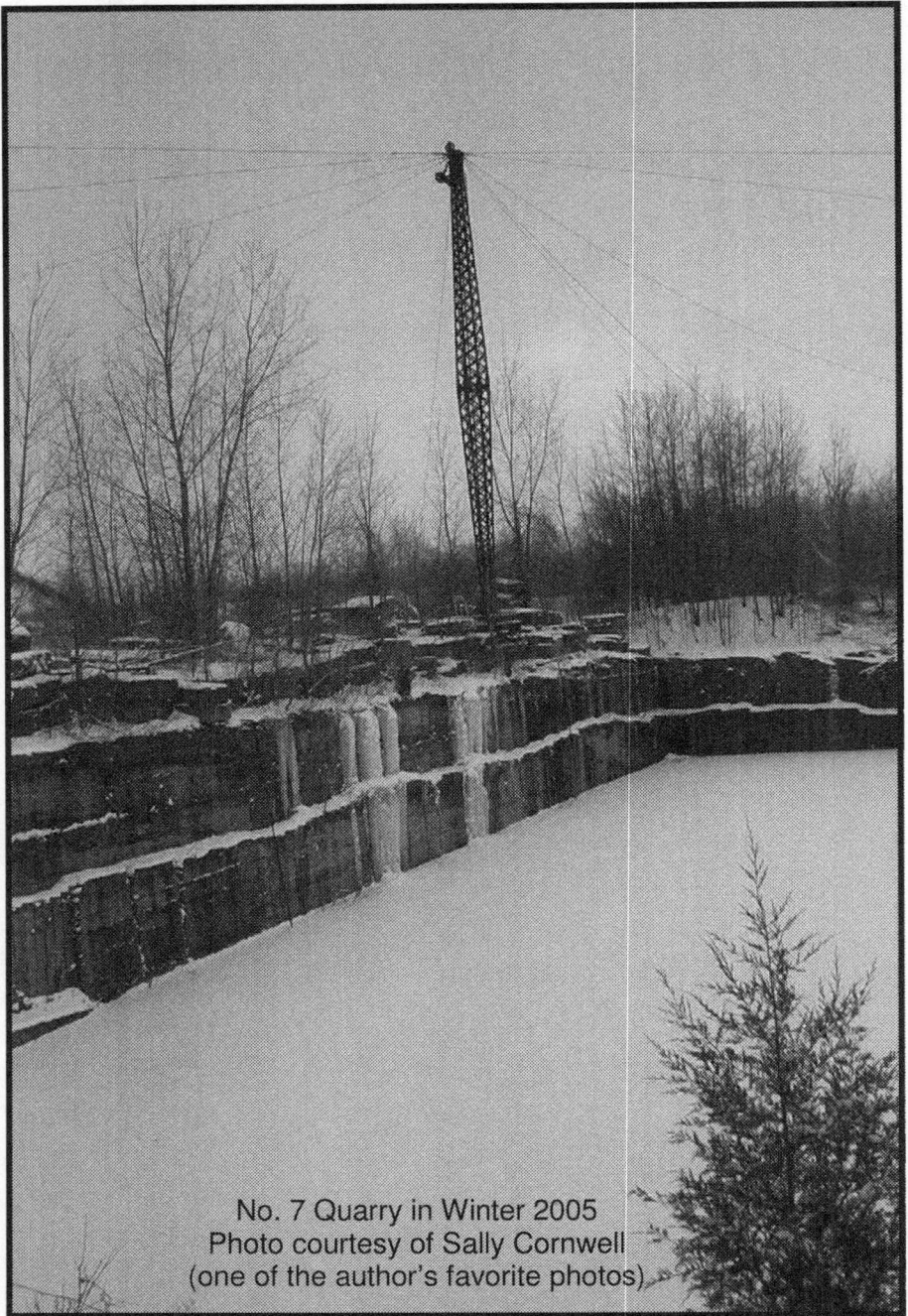

No. 7 Quarry in Winter 2005
Photo courtesy of Sally Cornwell
(one of the author's favorite photos)

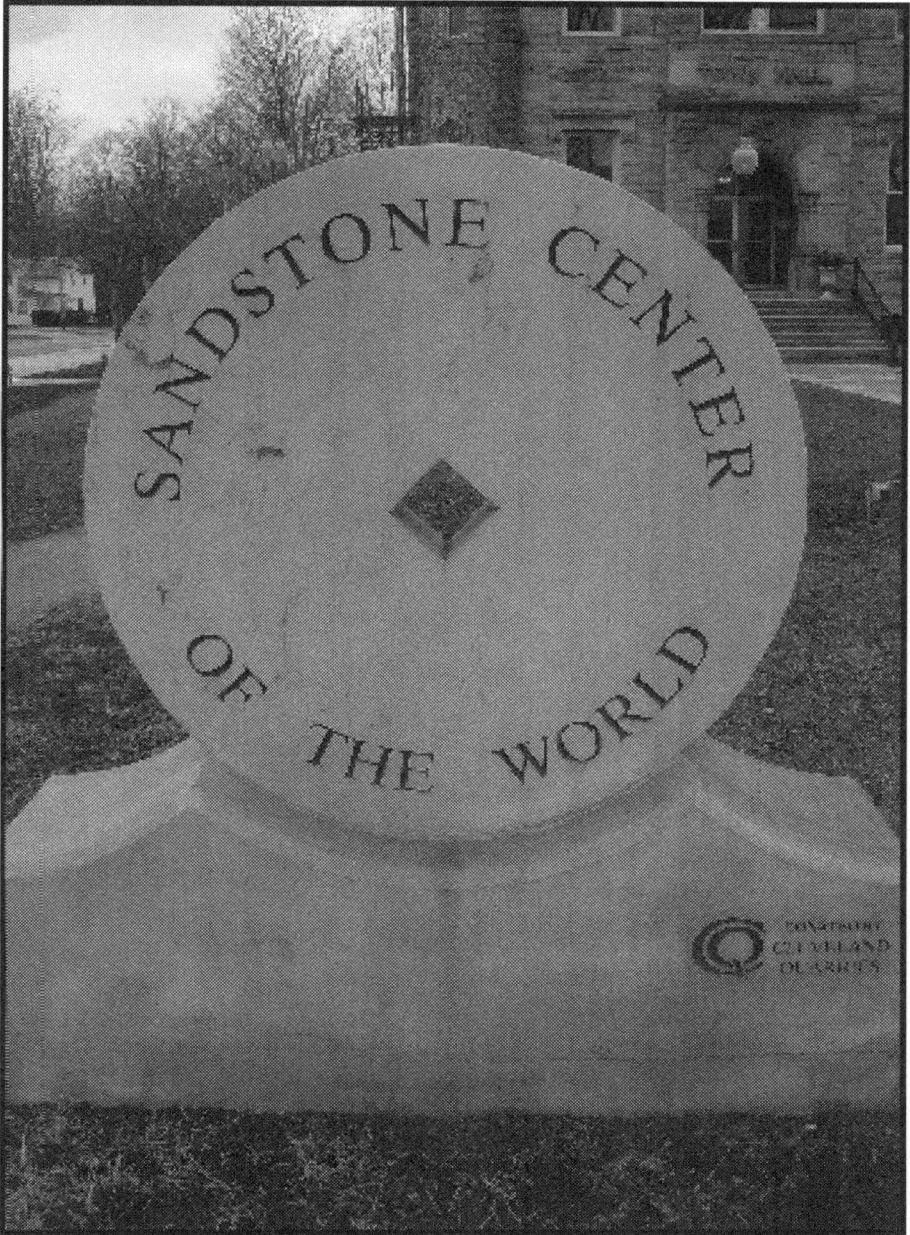

This stone marker resides outside the Amherst City Hall. It was donated by the Cleveland Quarries (Jim Hieb collection).

WWW.QUARRYTOWN.NET

www.ingramcontent.com/pod-product-compliance
Lightning Source LLC
Chambersburg PA
CBHW032104080426
42733CB00006B/418